AQUATIC HABITATS:

Exploring Desktop Ponds

Teacher's Guide

Grades 2–6

Skills
Observing, Comparing, Describing, Estimating, Measuring,
Communicating, Organizing, Experimenting,
Recording, Classifying, Drawing Conclusions

Concepts
Habitat, Food Web, Life Cycle, Adaptation, Decomposers,
Interdependence, Animal Structures and Behavior,
Biological Control, Environmental Characteristics

Themes
Systems and Interactions, Patterns of Change,
Structure, Energy, Matter

Mathematics Strands
Number, Measurement, Statistics

Nature of Science and Mathematics
Cooperative Efforts, Real-Life Applications, Interdisciplinary

by

Katharine Barrett and Carolyn Willard

LHS GEMS

Great Explorations in Math and Science
Lawrence Hall of Science
University of California at Berkeley

Design and Illustrations
Lisa Klofkorn

Cover Design
Lisa Klofkorn

Photographs
Richard Hoyt
Laurence Bradley

Lawrence Hall of Science, University of California,
Berkeley, CA 94720-5200

Chairman: Glenn T. Seaborg
Director: Ian Carmichael

Initial support for the origination and publication of the GEMS series was provided by the A.W. Mellon Foundation and the Carnegie Corporation of New York. Under a grant from the National Science Foundation, GEMS Leader's Workshops have been held across the country. GEMS has also received support from: the McDonnell-Douglas Foundation and the McDonnell-Douglas Employee's Community Fund; Employees Community Fund of Boeing California and the Boeing Corporation; the Hewlett Packard Company; the people at Chevron USA; the William K. Holt Foundation; Join Hands, the Health and Safety Educational Alliance; the Microscopy Society of America (MSA); the Shell Oil Company Foundation; and the Crail-Johnson Foundation. GEMS also gratefully acknowledges the contribution of word processing equipment from Apple Computer, Inc. This support does not imply responsibility for statements or views expressed in publications of the GEMS program. For further information on GEMS leadership opportunities, or to receive a catalog and the *GEMS Network News*, please contact GEMS at the address and phone number below. We also welcome letters to the *GEMS Network News*.

International Standard Book Number: 0-924886-01-3

COMMENTS WELCOME !

Great Explorations in Math and Science (GEMS) is an ongoing curriculum development project. GEMS guides are revised periodically, to incorporate teacher comments and new approaches. We welcome your criticisms, suggestions, helpful hints, and any anecdotes about your experience presenting GEMS activities. Your suggestions will be reviewed each time a GEMS guide is revised. Please send your comments to: GEMS Revisions, c/o Lawrence Hall of Science, University of California, Berkeley, CA 94720-5200. The phone number is (510) 642-7771 and the fax number is (510) 643-0309. You can also reach us by e-mail at gems@uclink.berkeley.edu or visit our web site at www.lhs.berkeley.edu/GEMS.

Great Explorations in Math and Science (GEMS) Program

The Lawrence Hall of Science (LHS) is a public science center on the University of California at Berkeley campus. LHS offers a full program of activities for the public, including workshops and classes, exhibits, films, lectures, and special events. LHS is also a center for teacher education and curriculum research and development.

Over the years, LHS staff have developed a multitude of activities, assembly programs, classes, and interactive exhibits. These programs have proven to be successful at the Hall and should be useful to schools, other science centers, museums, and community groups. A number of these guided-discovery activities have been published under the Great Explorations in Math and Science (GEMS) title, after an extensive refinement and adaptation process that includes classroom testing of trial versions, modifications to ensure the use of easy-to-obtain materials, with carefully written and edited step-by-step instructions and background information to allow presentation by teachers without special background in mathematics or science.

Staff

Principal Investigator: Glenn T. Seaborg
Director: Jacqueline Barber
Associate Director: Kimi Hosoume
Associate Director/Principal Editor: Lincoln Bergman
Science Curriculum Specialist: Cary Sneider
Mathematics Curriculum Specialist: Jaine Kopp
GEMS Network Director: Carolyn Willard
GEMS Workshop Coordinator: Laura Tucker
Staff Development Specialists: Lynn Barakos, Katharine Barrett, Kevin Beals, Ellen Blinderman, Beatrice Boffen, Gigi Dornfest, John Erickson, Stan Fukunaga, Philip Gonsalves, Linda Lipner, Karen Ostlund, Debra Sutter
Financial Assistant: Alice Olivier
Distribution Coordinator: Karen Milligan

Workshop Administrator: Terry Cort
Materials Manager: Vivian Tong
Distribution Representative: Felicia Roston
Shipping Assistants: Jodi Harskamp, Patrick Trombley
GEMS Marketing and Promotion Director: Gerri Ginsburg
Marketing Representative: Matthew Osborn
Senior Editor: Carl Babcock
Editor: Florence Stone
Principal Publications Coordinator: Kay Fairwell
Art Director: Lisa Haderlie Baker
Senior Artist: Lisa Klofkorn
Designers: Carol Bevilacqua, Rose Craig
Staff Assistants: Larry Gates, Trina Huynh, Chastity Pérez, Dorian Traube

Contributing Authors

Jacqueline Barber
Katharine Barrett
Kevin Beals
Lincoln Bergman
Susan Brady
Beverly Braxton
Kevin Cuff

Linda De Lucchi
Gigi Dornfest
Jean Echols
John Erickson
Philip Gonsalves
Jan M. Goodman
Alan Gould

Catherine Halversen
Kimi Hosoume
Susan Jagoda
Jaine Kopp
Linda Lipner
Larry Malone
Cary I. Sneider

Craig Strang
Debra Sutter
Herbert Thier
Jennifer Meux White
Carolyn Willard

Reviewers

We would like to thank the following educators who reviewed, tested, or coordinated the reviewing of *Aquatic Habitats* and *Dry Ice Investigations*. Their critical comments and recommendations, based on classroom and schoolwide presentation of these activities nationwide, contributed significantly to this GEMS publication. Their participation in this review process does not necessarily imply endorsement of the GEMS program or responsibility for statements or views expressed. Their role is an invaluable one; feedback is carefully recorded and integrated as appropriate into the publications. **THANK YOU!**

CALIFORNIA

Albany Middle School, Albany
Joanna K. Pace

Longfellow Middle School, Berkeley
Crispin Barrere
Betty H. Merritt
*Susan Tanisawa

Oxford School, Berkeley
Joe Brulenski
Barbara Edwards
*Janet Levenson
Anne Prozan

School of the Madeleine, Berkeley
Barbara Basinet
Tom Dwyer
Heather Skinner
*Judy Velardi

Stanley Intermediate School, Lafayette
Glenn Hoxie
*Michael Meneghetti
Michael Merrick
Dixie Mohan

Claremont Middle School, Oakland
Susan Cristancho
*Malia Dinell
Courtney Terry
Ann Tingley

Hawthorne School, Oakland
Doug Dohrer
Sonja Ebel
Sonny Kim
*Madeline Lee
Rebecca Marquez

Montclair Elementary School, Oakland
Joe Danielson
Margaret Dunlap
Terry Anne Saugstad
Sheila Sims
Michael Strange
*Sharon Tom

Sherman Elementary School, Oakland
Andrea Mitchell
Marty Price
Jean Rains
*Linda Rogers

Cinnabar School, Petaluma
Judy Bowser
Diana DeMarco

Liberty School, Petaluma
*Fran Korb
Preston Paull

Coronado Elementary School, Richmond
*Jody Anderson
Kevin Eastman
Heidi Garcia
Norah Moore

Edendale Elementary School, San Lorenzo
Terri Kaneko
*Cheryl Tekawa-Pon
Alison Williams

Glen Cove School, Vallejo
Marcia Burnham
Linda Combs
*Cindy Jones
Bruce McDevitt

COLORADO

Cory Elementary, Denver
Carol Calkin
Kay DeLong
*Scott Sala
Debbie Stricker
Margaret Wing

Paonia Middle School, Paonia
Kimber Arsenault
Becky Ruby
*Morrie Rupp
Larry Thompson

ILLINOIS

Williamsville Junior High School, Williamsville
Rhonda Fulks
*Marcie Lane
Julie McPherson
Polly Wise

MASSACHUSETTS

Westfield Middle School, Westfield
Marsha Estelle
Sue Regensberger
*Lisa Strycharz
Sybil Williams

NEW MEXICO

Bernalillo School, Bernalillo
*Belinda Casto-Landolt
Kim Chavez
Sue Fleming
Michelle Mora
Susan Rinaldi
Teresa Whitehead

NEW YORK

Makowski School, Buffalo
Pat Benson
*Sandy Campbell
Sarah Johnson
Dennis Knipfing
Caroline Parrinello

PENNSYLVANIA

Sewickley Academy, Sewickley
Bernice Boyle
*Vicki Carbone
Susan Harrison
Dolly Paul
Lori Sherry

TEXAS

Armond Bayou Elementary, Houston
Nancy Butler
Jennifer Chiles
*Myra Luciano
John Ristvey
Lessa Young

Murry Fly Elementary, Odessa
Joy Ellison
Ruben Evaro
Sandra Galindo
*Kym Monacelli
Maria Torres

UTAH

Cook Elementary, Syracuse
*Laurel Bain
Audrey Francis
Natalie Merrill
Kristi Shelley
Liesa Tobler

WASHINGTON

Challenger Elementary, Issaquah
*Roberta Andresen
Kathy Caravano
Bonnie Cole
Joyce Neufeld

* Trial test coordinators

Acknowledgments

A special thanks to Cathleen Justino and her wonderful third grade students at John Muir Elementary School in Berkeley. They enthusiastically tested the pilot activities and contributed important ideas, modifications, and illustrations. The many other teachers and students who helped us test and refine this guide are acknowledged separately in the "Reviewers" section.

The Alameda County Mosquito Abatement District provided a wealth of information and advice. Patrick Turney and Lee Holt delivered the mosquito fish used in the local trials of this teacher's guide, and Everett King, Environmental Specialist, provided a great deal of help in developing the list of sources for Gambusia and the list of mosquito-related Internet sites.

Vincent Resh, Professor of Insect Biology at the University of California at Berkeley, provided excellent scientific review and offered many invaluable suggestions.

Ted Robertson of the LHS and Kimi Hosoume, GEMS Associate Director, provided inspiration, guidance, and many valuable suggestions for additions to the activities and the background information. Laura Tucker, GEMS Workshop Coordinator and veteran environmental educator, provided much valuable input on many aspects of the guide, especially the aquatic field trip in Activity 5.

Many thanks go to Christine Bartlett and Pablo Ibañez, biology instructors at LHS, for presenting the unit to their summer camp participants. This provided the opportunity to obtain the wonderful photos that appear throughout the guide.

Some illustrations have been reprinted with permission from the Lawrence Hall of Science's Outdoor Biology Instructional Strategies (OBIS), copyright 1982 by The Regents of the University of California. In addition, a selection of the activities recommended for the aquatic field trip in Activity 5 are adapted and modified from the OBIS program.

Contents

Introduction

Aquatic Habitats is designed to provide students with a highly motivating and unique opportunity to investigate an aquatic habitat. As students set up and add organisms to their "desktop ponds," they become drawn into the compelling drama of life, in all its complex and cyclical vitality.

As students learn about the adaptations and interactions of organisms within the aquatic habitat, they gain essential understandings in life science and develop important skills. Students build an understanding of biological concepts through direct experience with living things, their life cycles, and their habitats. Throughout *Aquatic Habitats* students directly investigate the structures and behaviors of each pond organism, observing how the animal's adaptations help it move, find food, and escape predators. As they witness the life cycle of the mosquito, students discover that this familiar pest is eaten by fish. Students document and share their predictions and observations in every part of the unit.

There is a wonderfully fascinating quality about underwater worlds. Aquatic life exhibits amazing structures and behaviors. Unlike land organisms, aquatic organisms can "breathe" under water; they can use strange paddles and jets to maneuver. Their protective adaptations range from tough armor and spines to delicate flotation balloons and "jelly" for protecting eggs.

Unlike the clean, filtered aquarium some students may have at home, in this unit the containers are transformed into model pond habitats. As in an actual pond, organic waste drifts to the bottom, creating an environment for growth of beneficial bacteria and algae. Over the weeks of the unit, students study and add one type of organism at a time—plants, worms, snails, fish, and mosquito larvae. They discover firsthand some of the complex interactions within a typical pond ecosystem. Each team's small tank becomes a laboratory for observing nature's processes.

We highly recommend that each student keep an ongoing journal of their observations, investigations, conclusions, and questions. You can decide on the best format for your class. Providing students with repeated opportunities to express their ideas through writing, drawing, group reports, and in class discussions is a crucial part of the learning process. In this unit, keeping a journal can also help students better comprehend the changes they are observing over time and perceive the interconnectedness of the pond

It is important to point out that the living organisms that make this unit so extremely fascinating for students do require some special attention on the part of the teacher. Please see "Important Notes on Preparation, Cost, and Scheduling" on page 4, the "Getting Ready" section for the whole unit and each activity, and the "Some Sources for Key Items" section for tips on gathering materials and organisms with a minimum of preparation time and expense.

ecosystem. Their observations and detailed descriptions can deepen their own learning and in turn lead to new questions for the class to discuss and investigate.

The investigations in *Aquatic Habitats* also help students gain insight into many of the big ideas of science that permeate not only the biological sciences, but all scientific disciplines, including systems and interactions, stability within systems, and patterns of change. What students learn from their study of model aquatic habitats contributes to a deeper understanding of real world ecosystems in their later education and throughout their lives.

The goals of this unit align well with the *National Science Education Standards* and other leading guidelines for excellence. As students gain experience with specific organisms and the aquatic ecosystem, they become better able to comprehend the diversity of organisms, their adaptations, and interdependence within an ecosystem. These concepts literally come to life in the model pond habitat your students investigate in this unit. Food webs, such as those students create in this unit, identify the relationships among producers, consumers, and decomposers. These experiences can provide the basis for students later on, in the 5–8 grade range, to focus more generally on populations (all individuals of a species that occur together in a given place and time) and ecosystems (all populations living together and the physical factors with which they interact).

If your students have had little prior experience investigating water, you may want to precede this unit with physical science activities that explore sinking, floating, and the water cycle.

These activities may be used as an introduction to the study of any aquatic system: pond, lake, stream, river, or seashore. There are always opportunities for further investigations in these ever-changing habitats. A culminating field trip to an aquatic environment provides an opportunity for students to apply what they have learned in the classroom to the diverse and complex natural environment.

Activity by Activity

In Activity 1, teams of four students set up classroom aquatic habitats in sturdy plastic tanks. They learn why you have removed chlorine from the water, then they add sand and gravel, water plants, and a shelter to the tanks. Students draw and describe their habitats and predict the changes they might see in the days to come.

In Activity 2, students observe tiny red burrowing creatures known as Tubifex worms, then observe the responses of the worms to the water environment. Next, students observe the structures and movements of water snails, then add them to the habitats. Students once again record their observations and predictions. This activity introduces the concept that all organisms have body structures and behaviors that have certain functions. In this context, fifth and sixth grade students learn about adaptations.

In Activity 3, pairs of students observe the structures of fish in small clear cups, before adding them to the habitats. After the teams observe and discuss the exciting interactions between the organisms in the habitats (the fish eat the Tubifex worms!), students record observations and make predictions.

In Activity 4, as the model habitats more and more begin to resemble an actual pond, the complex interactions and balances of an ecosystem begin to "come together" conceptually for your class. In the first part of Activity 4, each student observes mosquito larvae in a clear plastic cup. Some larvae are added to their aquatic habitat, and students observe the responses of the larvae and the fish (the fish eat the larvae). The remaining larvae are put in a covered class mosquito holding tank so the students can observe the development and metamorphosis of the larvae into pupae then into adult mosquitoes. Some teachers may prefer to make the second part of Activity 4 a separate class session. Students are encouraged to speculate on what eats the dead plant leaves and animal waste (Tubifex worms and bacteria). They learn the importance of decomposers. This activity includes options for presenting food webs for students at different grade levels.

Activity 5 offers further investigations in the classroom and/or on a pond field trip, with many suggested ways to deepen your students' understanding of habitat, food web, and adaptations. Many teachers have used *Aquatic Habitats* as a basis for a semester-long series of investigations, similar to the ongoing investigations suggested in a companion GEMS guide, *Terrarium Habitats*.

While we outline several possible directions for ongoing investigations, and suggest a number of possible activities should your class be able to make a field trip to a local pond or stream, it is likely that you and your students will come up with many great ideas on your own. We'd love to hear your ideas and we'd welcome samples of student work and anecdotes about your experiences with *Aquatic Habitats*.

Terrarium Habitats, *for Grades K–6, is a wonderful companion guide to* Aquatic Habitats. *Groups of students design and construct terrariums for the classroom. Sowbugs, earthworms, snails, and crickets are placed in the terrarium habitat and students observe and record changes over time. There are detailed instructions on setting up and maintaining the terrariums, along with concise and interesting biological information on a number of possible small organisms that can become terrarium inhabitants.*

Important Notes on Preparation, Cost, and Scheduling

Before beginning the unit, be sure to allow ample time—a month or more—to check sources for organisms and materials and to plan how and when the organisms will arrive in the classroom. Whatever your schedule, this has to be taken into account in your planning. This preparation can be quite time consuming the first time you teach this unit. Doing things at the last minute can result in extra expense and problematic delays. Once you have what you need, the preparation time for each activity is quite reasonable.

We strongly recommend that you read the guide through in advance, and enlist help in the advance preparation. As in many other activity-based science units, the preparation load can be lessened considerably if you collaborate with another teacher or teachers. Parent, grandparent, or other adult volunteers can also provide much assistance.

Budget is an important factor for you to consider, depending on what materials/organisms you may already have available and the resourcefulness of you and your school or district. (We know GEMS teachers are resourceful!) *If* you had to mail order all the materials and organisms for one class (including the tanks used for the habitats) it would cost about $200. Once you have the tanks, sand, and gravel, it could still cost about $150 for subsequent classes if you had to mail order all the organisms. A great deal of this cost can be reduced in ways we suggest in "Getting Ready for the Whole Unit" and "Some Sources for Key Items." The more able you are to obtain items through donation and other alternative methods and sources, the lower the costs will be!

We emphasize these considerations to make sure you are aware of them from the start, not in any way to discourage you from presenting the activities! The many teachers who tested these activities strongly affirmed that the educational impact of this unit was well worth it. The plants and animals chosen for these activities should be quite accessible to all teachers and the expenses reasonable given the high return in long-term learning!

As always, we welcome your ideas and suggestions. If you are able to find good sources for organisms or materials used in *Aquatic Habitats*, write a letter to the *GEMS Network News*, call, or send us an e-mail. GEMS guides are revised frequently based on continuing teacher feedback and we welcome your comments!

Time Frame

The five main activities in the unit can be scheduled over a few weeks, or can extend over several months. It's desirable to allow at least several days between Activities 1 and 2. This gives algae and beneficial bacteria time to become established, and allows chemical imbalances in the water to stabilize. Try to build flexibility into your teaching schedule so you can adapt to the arrival of organisms.

Mosquito larvae (Activity 4) are easier to get and keep alive during warmer seasons. The best time for a field trip (Activity 5) is during early fall or spring when insects and other small animals are abundant in and around ponds and streams.

Activity 1: Creating an Aquatic Habitat .... 30–60 minutes

Activity 2: Tubifex Worms and Snails .... 45–60 minutes

Activity 3: Fish Enter the Habitat .... 45–60 minutes

Activity 4: Mosquitoes .... one 60-minute or
two 45-minute sessions

Activity 5: Further Explorations

Part I: Ongoing Investigations with Classroom Aquatic Habitats .... 20 minutes/day

Part II: An Aquatic Field Trip .... several hours

Getting Ready for the Whole Unit

The availability and cost of organisms and certain materials can vary greatly by region and time of year. We recommend that you or a volunteer invest the time needed before beginning the unit to line up what you'll need, focusing especially on the nine key items listed in the following chart. This way, you'll avoid unexpected expenses or delays, and be better able to enjoy the magic of the developing habitats along with your students.

Here are some tips to help you prepare for the unit:

- Read over the guide in advance.

- Jot down on your calendar about when you'll need the materials and organisms for each of the five main activities.

- Make telephone calls or visit local stores to find your best sources, the cost, and availability.

- Order each organism or arrange for pick up at the right time in the unit. To avoid "die offs" of worms, snails, fish, or mosquito larvae, schedule each order carefully to arrive in your classroom within a day or two of when you're ready to use the organisms with students.

If you find one or more of the recommended items is hard to find or too expensive in your area, try one of the alternatives listed below. **For additional information, see "Some Sources for Key Items" on page 88.**

Check in your area about the availability of all of the organisms. In most aquarium stores, for instance, the recommended Elodea plants are available and inexpensive. However, teachers in Washington state found that Elodea is illegal in their state because it has been a problem in clogging wild waterways. They chose to use another water plant. We'd welcome hearing of other sources and additional alternatives.

Recommended Item (approximate amounts per class)	Good Alternatives	Notes

Activity 1

8 aquarium containers with lids 1 ½ gallon rectangular, clear, flexible plastic tanks	Ask parents to save clear, sturdy, squat containers that some bulk foods are packaged in. (Bulk popcorn kernels, licorice, and other items are sometimes found in such containers.)	The sides of the containers should be easy to see though without any distortions. Lids could be fashioned from flexible plastic screen. Holes must be ¼" in diameter or smaller to prevent escape of adult mosquitoes.
10 lbs. gravel white or light-colored aquarium gravel	Any light-colored gravel, but not coral gravel which is used only for salt water tanks.	If you can't get gravel, you can substitute sand; in that case, habitats will just have one type of bottom covering.
10 lbs. sand white or light-colored aquarium sand	Any light-colored sand, not too fine or dusty. (Don't use coral sand which is used only for salt water tanks.) Try coarse gardening sand, good quality sandbox sand, or sand-blasting sand.	See page 15 for information on rinsing dusty sand. If you can't get sand, you can substitute gravel; in that case, habitats will just have one type of bottom covering.
dechlorinated water tap water and dechlorinating liquid	Bottled spring water is also fine. Don't use distilled water because it lacks beneficial minerals.	In *some* regions, water can be dechlorinated simply by leaving it in a container without a lid for about 24 hours. See page 15 for more about treating water.
8 or more sprigs of water plant Elodea plant	hornwort anacharis peacock feather plant duck weed	Any green freshwater plant, preferably one that snails can eat.

Recommended Item (approximate amounts per class)	Good Alternatives	Notes
Activity 2		
small organisms that fish can eat an ounce of Tubifex worms, also known as "blood worms"	Daphnia or any other small pond crustacean	Note: brine shrimp are often fed to fish, but they die soon if not eaten and aren't native to fresh water.
16 or more water snails small, pond snails	If your local aquarium store doesn't carry pond snails, you may need to buy the larger and more expensive aquarium snails, in which case you may opt to buy just one per tank.	
Activity 3		
about 3 dozen fish Gambusia (also known as mosquito fish)	Small "feeder" goldfish; hardy "feeder" guppies; freshwater minnows.	Avoid fish that need heated water. Fish should be between about ½" and 1 ½" long. Catching small fish such as minnows in a local pond can work. **See information on page 91 about getting FREE Gambusia from mosquito abatement or vector control agencies.**
Activity 4		
mosquito larvae order larvae from a scientific supply company (see "Some Sources for Key Items" on page 88)		If you want to start your own mosquito culture in a bucket (see page 84), begin several months before this activity. Mosquito larvae could also be obtained from mosquito abatement or vector control agencies.

What You Need for the Whole Unit

The quantities below are based on a class size of 32 students. Depending on the number of students in your class, you may, of course, need fewer materials.

This list gives you a concise "shopping list" for the entire unit. Please refer to the "Getting Ready" sections for each activity. They contain more specific information about the materials needed for the class and for each team of students.

Non-Consumables

- ❑ a 3–5 gallon container for a class "holding tank"
- ❑ a bucket or other large container
- ❑ 8 1 ½ gallon plastic flex-tank aquariums and 8 lids with small, round holes (about ¼" in diameter)
- ❑ 32 clear plastic cups
- ❑ 8 shelters (cups, strawberry baskets, or plastic flower pots)
- ❑ one or two cottage cheese-type containers with perforated lids
- ❑ a turkey baster or large syringe for transferring worms and mosquito larvae/pupae (the worms will get stuck in a medicine dropper)
- ❑ one small aquarium net
- ❑ a wide-mouthed container for a mosquito holding tank
- ❑ a 12" tube of pantyhose
- ❑ several light-colored dishpans or clear, plastic sweater boxes to use as observation trays for large organisms found at the pond
- ❑ 8 dip nets or small (about 4" in diameter) metal strainers with handles
- ❑ 16 disposable, white cereal bowls or other light-colored basins to use as observation trays at the pond

Optional:
- ❑ an extra flex-tank aquarium and lid (plus sand, gravel, Elodea, shelter, label, snail, Tubifex worms, fish, and mosquito larvae) to use in class demonstrations
- ❑ 32 hand lenses

Consumables

- ❑ about 20 gallons of dechlorinated water
- ❑ about 10 pounds pea-sized, light-colored aquarium gravel
- ❑ about 10 pounds white or light-colored sand
- ❑ about 10–12 sprigs of Elodea or another water plant
- ❑ 1 bottle of dechlorinating liquid
- ❑ about 18–20 aquatic snails
- ❑ one ounce of Tubifex worms
- ❑ about 3 dozen Gambusia fish
- ❑ one container fish flake food
- ❑ about 50–100 mosquito eggs or larvae

Copies of the following:
- ❑ 8 sets of 4 Aquatic Habitats Task Cards (master on page 21)
- ❑ 128 Aquatic Habitats student sheets (master on page 22) *OR* 32 journals
- ❑ 32 Mosquito Life Cycle student sheets (master on page 56)
- ❑ several Guide to Freshwater Life handouts (master on pages 76–78)

Optional:
- ❑ 32 copies of the Parts of a Fish student sheet (master on page 43)
- ❑ 32 copies of the "Too Many Mosquitoes!" homework or class reading assignment (master on pages 57–59)

General Supplies

- ❑ 1 ruler
- ❑ 1 black permanent marker
- ❑ 1 pair of scissors or a paper cutter
- ❑ newspaper and/or paper towels
- ❑ several sheets of chart paper
- ❑ 8 adhesive labels
- ❑ 32 pencils
- ❑ 32 pieces of white scratch paper
- ❑ a large rubber band

Optional:
- ❑ 32 sheets of 18" x 9" construction paper for folder or journal covers
- ❑ crayons or markers for journal cover drawings
- ❑ a small dry-erase (white) board and marker for recording class observations at the pond

Activity 1: Creating an Aquatic Habitat

Overview

In this activity, students are invited to explore an underwater world through the windows of a model pond. The class identifies the necessary components of a habitat, and groups of four students work together to arrange their habitats to receive animals in later activities.

Student groups add clean sand and gravel, plants, and a shelter to their tanks of dechlorinated water. They then draw and describe their aquatic habitats and predict the changes they might see in the days to come.

Please note: Contrary to the usual goal for a home aquarium, you *want* algae to grow in these model pond habitats. Ideally, after the aquatic habitats have been established for a while, green algae will be visible growing on surfaces and in the water.

What You Need

For the entire class:
- ❑ a 3–5 gallon container for a class "holding tank" (See "Getting Ready" for details.)
- ❑ a bucket or other large container to hold extra dechlorinated water
- ❑ about 10 pounds pea-sized, light-colored aquarium gravel
- ❑ about 10 pounds white or light-colored sand
- ❑ 1 ruler
- ❑ 1 black permanent marker
- ❑ 1 pair of scissors or a paper cutter
- ❑ a few extra sprigs of Elodea water plant for the holding tank
- ❑ 1 bottle of dechlorinating liquid
- ❑ newspaper and/or paper towels to mop up spills
- ❑ chart paper for recording questions
- ❑ (optional) 1 extra flex-tank aquarium with lid, label, gravel, sand, and shelter to use in class demonstrations

Note: Ten pounds each of sand and gravel is enough for eight teams of four students to have about one cup of each. If necessary, a half-cup each of sand and gravel per team is adequate. If possible, save a little to put in your class holding tank.

For each group of four students:
- ❏ a 1 ½ gallon plastic flex-tank aquarium and a lid with small, round holes (about ¼" in diameter)
- ❏ a sprig of Elodea or other water plant
- ❏ 2 clear plastic cups (one each for sand and gravel)
- ❏ dechlorinated tap water
- ❏ 1 shelter (cup, strawberry basket, or plastic flower pot)
- ❏ 1 adhesive label for group names
- ❏ 1 set of 4 Aquatic Habitats Task Cards (master on page 21)

For each student:
- ❏ a pencil
- ❏ 4 copies of the Aquatic Habitats student sheet (master on page 22) *OR* a journal
- ❏ *(optional)* an 18" x 9" sheet of construction paper to make a folder or journal cover
- ❏ *(optional)* crayons or markers for journal cover drawings

Getting Ready

Before the Day of the Activity

1. Decide how to assign students into groups of four for the *Aquatic Habitats* unit. If necessary, plan how to rearrange desks or tables.

2. Decide where the habitats will be kept during the unit. Most teachers have them at students' desks during the activities, and move them to a table or counter for the rest of the time. However, depending on grade level, the stability of desks or tables, the room arrangement, and other factors, many teachers find it a great advantage to allow teams to keep their habitats at their desks all day. It is not necessary to keep the habitats near a window—in fact direct sunlight may cause the water to overheat. Decide whether students will be allowed to move the tanks or whether you'll do this.

3. Purchase from an aquarium supply store:

- **Water plants:** Elodea (el-oh-dee-ah), or water weed, is a common, hardy, and inexpensive water plant available at most aquarium stores. You will need at least one sprig per team of students plus several sprigs for your holding tank. Sprigs are usually about 8"–12" long.

- **Sand and gravel:** Both sand and gravel should be light in color and free of dust and debris. Sand and/or gravel purchased from an aquarium store is usually clean. (Remember not to use coral sand or gravel, which is for salt-water tanks.) If you buy sand and gravel elsewhere, it could be less expensive, but you may need to rinse them a few times to remove excess dust. If rinsing is necessary, just place sand or gravel in a bucket, add water, stir, pour off grit and dust, and repeat until the water is clear.

Light-colored sand and gravel will allow students to more easily see the animals and the detritus (dee-TRY-tus)—dead leaves, feces, dead organisms, and other waste that falls to the bottom of their tanks. Later, they will learn that as detritus decomposes, essential nutrients are provided for other organisms.

- **Dechlorinating liquid:** In many parts of the country, tap water is now treated with a very stable form of chlorine called "chloramine" that does not dissipate. If this is the case in your region, you will need to use dechlorinating liquid. (In some regions, the type of chlorine in the tap water will evaporate if you simply leave the water in a container without a lid for about 24 hours. If this is the case in your area, you won't need dechlorinating liquid, unless you need to dechlorinate quickly.)

Check with your aquarium store to learn which type of chlorination is used with your local tap water, and, if necessary, purchase a dechlorinating liquid (often called a water conditioner). One drop of this handy liquid will instantly dechlorinate a gallon of water. There are a number of brands of dechlorinating liquid on the market.

4. Set up your holding tank. We recommend using a restaurant bus tub because it is shallow and provides a large water surface for aeration. Other options for the holding tank include a 3–5 gallon aquarium—or several smaller, watertight, plastic or glass containers, such as sweater boxes. Your holding tank doesn't need a lid.

Each time new organisms arrive in your holding tank, they will be a great attraction for students. If necessary, set a few ground rules for observation, so that the organisms (and your students) will be calm and safe. Keep fishnets, food, dechlorinating liquid, and other equipment away from the tank, and out of sight.

a. Choose a light, cool place for your holding tank, away from heaters and direct sunlight.

b. Fill the holding tank about two-thirds full with water. Shallow water provides better aeration, and fish will be less likely to leap over the sides.

c. Dechlorinate the water, either with a water conditioner or by letting the water sit out for 24 hours.

d. Spread some gravel at one end of the holding tank and some sand at the other end.

e. Float the Elodea plants in the water. *(Note: Although Elodea can survive in chlorinated water, dechlorinating the water allows for the growth of beneficial bacteria and algae in the holding tank, and prepares for the later arrival of snails and fish.)*

f. Fill an additional bucket or other large container with extra dechlorinated water.

g. When necessary, replace the water that has evaporated from the holding tank with dechlorinated water.

5. For each group of four students, prepare a 1 ½ gallon tank:

a. On one side of the tank, draw an about 2" long line about an inch down from the top using the ruler and permanent marker. This will be the water line.

b. Fill each tank to within ½" of the water line *before* the activity. (During the activity, students will add sand, gravel, shelter, and Elodea to the already filled tanks.)

c. Dechlorinate the water.

6. Collect 8–12 clean plastic containers to use as shelters in the aquariums. Small, lightweight, black or green plastic flower pots are ideal.

7. Duplicate four copies of the Aquatic Habitats student sheet (master on page 22) for each student, one for use in Activity 1, and the others for Activities 2–4. You could distribute one student sheet to each student during Activities 1–4, or you could staple four student sheets together—including blank pages—to make a journal for the whole unit. Teachers of younger students may want to draw the rectangular shape of the tank at the top of the student sheet before making copies.

8. If you prefer to use journals with blank pages rather than the Aquatic Habitats student sheets, decide on materials for your students to use in making their journals or folders. You may want to provide construction paper for covers, crayons or markers for cover drawings, etc. Select whatever format works best for students to keep a continuing record of their observations and ideas.

If you have only one sink, it will take quite a while to fill all the tanks. For this reason, we strongly recommend that you or a volunteer fill them before the classroom activity. Please note that a 1 ½ gallon container of water weighs about 12 pounds.

9. Make one copy of the Aquatic Habitats Task Cards sheet (master on page 21) for each student group, cut them apart, and stack them.

On the Day of the Activity

1. Fill a cup with sand and a cup with gravel for each team.

2. Arrange group materials (sand, gravel, Elodea, shelter, tank filled with water, and label) away from student work areas, in a place where students can easily gather them without crowding. Set out the Elodea as individual sprigs so that eager students don't inadvertently pull the plants apart.

3. Put newspapers or paper towels nearby in case of spills.

4. Have handy the sets of task cards for each team.

5. Prepare a piece of chart paper with the title "Aquatic Habitats Questions," and have it handy to put on the wall during the closing discussion.

6. Have the Aquatic Habitats student sheets or journals on hand. If you've decided to use them, provide construction paper and crayons or markers for students to make folders or journal covers, and decide whether you want to have students construct the journals before the unit or during the first activity.

Teachers of younger students may want to have an extra set of materials and extra 1 ½ gallon tank to use to introduce (model) each activity.

Introducing Aquatic Habitats

1. Start by asking, "What kinds of animals and plants live under water?" Accept any ocean or fresh water organisms as answers.

2. Show students one of the 1 ½ gallon containers of water and ask what might be able to live in it. If students say sharks and whales, point out the size of the tank. If they mention marine organisms, ask students how ocean water differs from pond water. [ocean water is salty] Tell them this is **not** salt water.

3. Once students have mentioned a variety of fresh water organisms, including fish, ask them to help you think of things a fish would need to live. List their ideas on the board. [clean water, food, light, hiding places, stuff on the bottom]

4. Write the word *habitat* on the board, and explain that a **habitat is a place that has everything an animal needs to live.** "Habitat" is a scientific name for a home. Write the word *aquatic* in front of habitat, and explain that it comes from the Latin word *aqua*, which means water.

5. Tell the students that today they will get to set up small aquatic habitats. In the next few weeks, they will get to add to the habitats some small animals that live in ponds and streams.

6. Explain that our tap water contains chlorine to kill germs. There usually isn't enough chlorine to bother humans, but it can hurt or kill small aquatic animals. Explain how you removed the chlorine from the water in their tanks.

Making an Aquatic Habitat

1. Explain that students will work in groups of four. Each group will set up and share one aquatic habitat. Put the students together in their groups of four and have them number off one to four. Explain that each group member will have a task that goes with their number.

2. Show the materials and explain the tasks that students will follow to make their tanks into aquatic habitats:

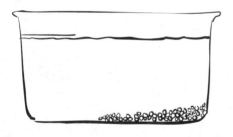

> **Task 1:** Get a cup of gravel. Gently spread it over one half of the tank.

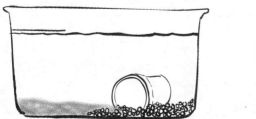

> **Task 2:** Get a cup of sand. Gently spread it over the other half of the tank, so that the gravel and sand touch but don't mix very much. (Point out that the placement of the gravel and sand will create two different kinds of bottom areas.)

> **Task 3:** Get a shelter. Talk with your team about where they want to put the shelter. Keep the shelter on the bottom by using some gravel or sand to weigh it down.

Task 4: Write your team's names in pencil on a label. Stick the label on the outside of the tank near the top. Get a sprig of Elodea plant and float it in the tank. (Let the students know that it's best not to try to "plant" the Elodea in the sand or gravel, because it usually comes out anyway.)

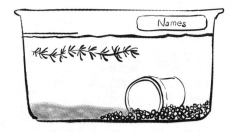

3. Pass out the task cards. If necessary, review the tasks. Say they should do the tasks in order, and cooperate with their teammates. Mention that if someone must put a hand in the water, it should be rinsed of all soap or lotion so chemicals don't pollute the water and then washed after the task is complete.

Some teachers pass out the task cards first, and have students follow along as they model the tasks.

4. Have students clear off their desks, and have each group member go get the material for their particular task while you deliver the tanks full of water and the tank lids.

5. Circulate among the groups, answering questions and encouraging discussion.

6. Once the habitats are set up, have the teams add lids. Collect empty cups, and mop up any spills.

Documenting the Habitat

Some teachers prefer to end the class session here and make the recording part of the activity a second class session.

1. Regain the attention of the class, and say that over the next few weeks their aquatic habitats will become "model ponds," and they will get to be scientists, studying things about pond life that people seldom see.

2. Point out that, as scientists, they need to keep good records of how their new habitat began. Hold up an Aquatic Habitats student sheet and let students know they'll all receive one of these.

3. Quickly sketch a sample of the sheet on the board. Demonstrate how to draw the outline of the tank at the top of the sheet, and add name and date.

4. Ask volunteers to describe their tanks. On the chalkboard, show how to draw and label what's being described. Under "Observations," write one or two of their observations.

5. Next encourage students to speculate on how their habitats might change over the next day or two, before

Because of their experiences with home aquaria, some students may ask if they'll need to clean their tanks. If they ask, explain that they will not need to be cleaned, as these are to be model ponds and the idea is to let them develop as they would in nature.

Many of these questions may be answered as a result of students' own observations. Some questions can later become topics for library research or investigations.

they add any animals. Accept all predictions. [water might become clearer, bubbles might form, the plant might drift, water might evaporate] Write one or two predictions on the chalkboard sample sheet under the heading "Predictions."

6. Let students know that **observations** tell what the tank is like now, and **predictions** are guesses about what the tank might be like in a day or two. Ask students to write all the observations and predictions they can on their own papers.

7. Distribute the Aquatic Habitats student sheets or journals.

8. If time and student interest permit, announce that you will start a class list of questions that come up during the investigations. Invite students to suggest three or four questions they have wondered about and record these on the chart paper.

Going Further

Set up a schedule for students to take turns over the next few weeks testing the chemical condition of the water with pH and ammonia test kits, which are available from aquarium stores.

Aquatic Habitat Task Cards

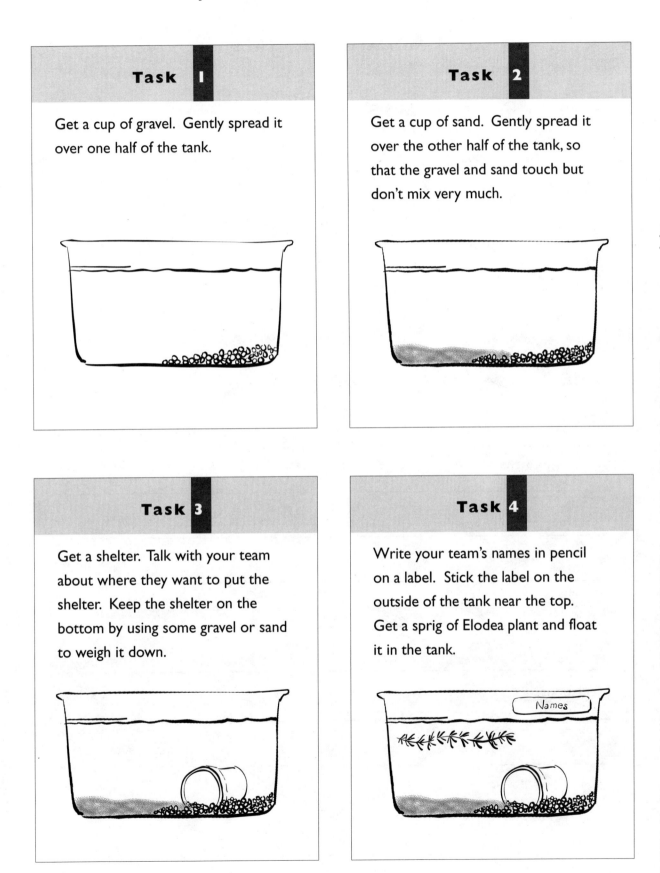

Task 1

Get a cup of gravel. Gently spread it over one half of the tank.

Task 2

Get a cup of sand. Gently spread it over the other half of the tank, so that the gravel and sand touch but don't mix very much.

Task 3

Get a shelter. Talk with your team about where they want to put the shelter. Keep the shelter on the bottom by using some gravel or sand to weigh it down.

Task 4

Write your team's names in pencil on a label. Stick the label on the outside of the tank near the top. Get a sprig of Elodea plant and float it in the tank.

Names

Date _____   Name _____

AQUATIC HABITATS

Observations:

_____

_____

_____

_____

_____

Predictions:

_____

_____

_____

_____

Activity 2: Tubifex Worms and Snails

Overview

In this activity, groups of students observe Tubifex worms, add the worms to the water habitat, and observe their behavior. Tubifex (TOO-bih-fex) worms are slender, segmented, reddish worms about an inch and a half long (four centimeters), that live in the bottom of ponds and slow moving rivers and streams. They construct tubes in the mud, out of which only the hind part of their bodies extend. Clumps of Tubifex worms sometimes look like patches of reddish, waving fringes on the bottom of quiet pools. **Don't tell your students yet, but fish eat Tubifex worms!** (They'll discover this for themselves in Activity 3.)

Like their terrestrial relatives, the earthworms, Tubifex worms depend on decaying matter such as dead leaves and animal waste. These harmless worms are recyclers, constantly consuming the organic matter that would otherwise fill the habitats, and transforming it into useful nutrients for plants. They also move bottom sediments to the surface by their movements.

Next, students observe the structures and behaviors of water snails and add them to their habitats. These creatures exhibit amazing forms of locomotion—some snails float, others "inch" along the surface of the water, and others glide across the walls of the tanks. Students may observe snails grazing on the Elodea and algae, and may also see snails mating and laying eggs.

This activity introduces the concept that all organisms have body structures that have certain functions. Fifth and sixth grade students learn that the behaviors and body structures that help an animal survive in a habitat are called *adaptations.*

If Tubifex worms are not available in your local aquarium supply store, or if your mail order worms die off in great numbers, don't despair! You can substitute Daphnia or other fresh water crustaceans that fish eat.

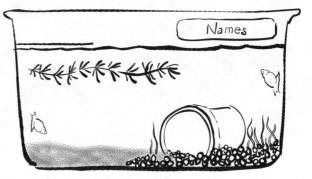

What You Need

For the entire class:
- ❏ 3–5 gallon holding tank from previous activity
- ❏ bucket of dechlorinated water from previous activity
- ❏ about 18–20 aquatic snails
- ❏ one ounce of Tubifex worms (See "Getting Ready" for more information.)
- ❏ one or two cottage cheese-type containers with perforated lids for holding Tubifex worms
- ❏ a turkey baster or large syringe for transferring worms (the worms will get stuck in a medicine dropper)
- ❏ the Aquatic Habitats Questions on chart paper from previous activity
- ❏ newspaper and/or paper towels
- ❏ *(optional)* an extra aquatic habitat from previous activity to use in class demonstration

For each group of four students:
- ❏ aquatic habitat from previous activity

For each pair of students:
- ❏ 2 clear plastic cups (one for about 20 Tubifex worms, one for a snail)
- ❏ a piece of white scratch paper
- ❏ *(optional)* hand lens

For each student:
- ❏ a pencil
- ❏ a new Aquatic Habitats student sheet (master on page 22) or journal

Getting Ready

Before the Day of the Activity

1. Obtain enough snails so that each pair of students will have one, plus some extras, and keep them in the holding tank with the extra plants. There are several species of freshwater (aquatic) snails sold in aquarium stores. Small pond snails are least expensive, if available, but any fresh-water snails are fine to use.

2. As close to the day of the activity as possible, purchase about one ounce (several hundred) of Tubifex worms from

your aquarium store or by mail order (see "Some Sources for Key Items" on page 88). Keep the worms in one or more cottage cheese-type containers with perforated lids, in **very shallow,** (about 1") **dechlorinated** water. The water should be just barely above the worms. The worms should survive for a day or two. Keep in mind:

- Shallow water provides a greater surface area, allowing oxygen from the air to diffuse into the water, replacing the oxygen that is drawn from the water by the worms.

- If the water becomes cloudy and an odor develops, replace some of the water with fresh, dechlorinated water.

- Do not put sand or gravel in the container. This will make it easier for you to use a turkey baster to remove the worms.

- If it's convenient, keep the worms in the warmest part of a refrigerator (usually in the door) before the day of the activity to decrease their metabolism and reduce odors. Check to make sure the worms don't freeze.

Tubifex worms often "clump" together in the container and look inactive—this is not necessarily cause to worry. To make sure they're still alive, put a few into your holding tank and see if they "unclump" and move around.

3. If you decide to use hand lenses, and your students have had little experience with them, you may want to plan an additional class session before this activity to let them examine a page of print (and their clothes, skin, desktops, and whatever else is handy). Encourage free exploration and help students learn to focus by moving the lens back and forth between the object and their eyes.

4. Make a copy of the Aquatic Habitats student sheet (master on page 22) for each student or use folders or journals prepared earlier.

On the Day of the Activity

1. Before class, prepare four clear plastic cups for each group. Each *pair* of students in the group will receive a cup of Tubifex worms and, later, a cup containing a snail.

a. **Tubifex worm cups:** add about an inch of dechlorinated water to the cups, then use the turkey baster to transfer about 20 Tubifex worms from the cottage cheese container to each cup. Since the worms tend to clump when disturbed, you may want to separate them

gently with the tip of the baster before transferring them.

b. **Don't worry about getting an exact amount of worms in each cup.** The habitats can comfortably accommodate quite a few worms. However, avoid adding too many **dead** worms or lots of cloudy, bad-smelling water to the cups. (If added to the habitats, this can cause a big increase in the population of bacteria, which use much of the oxygen that the other organisms will need.) So, if many of your worms have died, it's worth taking some extra time to try to sort out mostly the living ones for transfer to the students' cups.

c. **Snail cups:** add about an inch of dechlorinated water to each cup, then add one snail.

d. **Cover the cups** with newspaper or otherwise keep them out of sight to avoid having students crowding around to see the animals.

2. Have a sheet of white scratch paper and, if you've decided to use them, a hand lens ready for each pair of students.

Introducing the Activity

1. If you collected the tanks after the previous activity, distribute them to the teams of students.

2. Invite several students to share what they have noticed in their model habitats. If you haven't already started a class list of Aquatic Habitat Questions, this may be a good time to introduce it.

3. Remind everyone that aquatic animals can be injured by chemicals such as chlorine and pollution in the water. Ask how they can make sure aquatic animals remain healthy. Mention the following "Habitat Health Rules" if they aren't mentioned by students (they could even be posted):

If students have noticed that the water level in their habitats has gone down, allow them to fill them to the water line with dechlorinated water later in this activity, and in future activities.

• Tap water added to the tanks must be dechlorinated.

• Keep hands out of water. If someone must put a hand in the water, it should first be rinsed of all soap or lotion so chemicals don't pollute the water. After being in the tank, hands should be thoroughly washed.

- Don't hold the animals.

- Add the animals gently to the water.

- Try not to scare the animals; use quiet observation skills.

Observing Tubifex Worms

1. Tell students they will first be observing a very common animal that is similar to a tiny earthworm except that it lives on the bottom of ponds. Write the name "Tubifex Worm" on the board. Display a cup of worms for students to see and let them know that pairs of students will each receive a cup of worms.

2. Sidestep any negative comments at this point with an encouraging comment that these small harmless creatures have an important role to play in the model habitats.

3. Make clear that partners will **not put their worms into their tanks right away,** but should first put the cup on the table where both students can easily see it. Holding the white scratch paper behind the cup sometimes makes it easier to see the worms. Ask students to observe the worms and try to estimate how many are in the cup. How do the worms move? Students could sketch the worms. Do they notice any color variations? Encourage specific observations.

4. Distribute the cups and scratch paper. Circulate, reminding students to use their "quiet observer" skills and not to release the worms into the tanks yet.

5. Regain the attention of the class. Have volunteers share their observations. Ask students to speculate on why the worms clump together. Encourage predictions about what will happen when the worms are added to the habitat.

6. Use an empty cup to model how students should gently lower the cup into the water so all the worms can swim out into the tank.

7. Tell students that one pair of students will release their worms while their whole team observes the behavior of the worms. After a few minutes, the second pair may release their worms.

One teacher said that upon receiving the worms some of her students "loved to act disgusted, but soon got over it."

Students will probably observe the worms burrowing in the sand or gravel. After a while, it may seem to students that all of their worms are gone! Have them look very closely to see if some of them are sticking up out of the sand or gravel. If necessary, lift the tanks for the students to see if worms are visible from the bottom.

Maerose 2/5/97

Discussing Observations and Introducing Water Snails

1. After they have had a few minutes to observe, regain the attention of the class, and invite several volunteers to share worm observations and questions.

2. Ask questions, such as: How did the worms move? Do the worms seem to prefer the sand or the gravel bottom (also called "substrate")? What do you think the worms might eat? [plants, other animals, each other, dead things] Point out that students may be able to discover the answers to these questions by observing the worms in the model habitats.

3. Announce that it's time to observe another animal that lives in most aquatic habitats around the world. Show students the snails in cups and invite predictions about what the snails might do when released in the habitat.

4. Challenge students to observe the snails in the cups first, then release them into their tanks. **Make sure that students understand that, unlike land snails, these snails need to stay in the water to survive.**

5. Distribute the cups and ask students to observe the snails. Circulate among the teams listening to observations and asking questions such as: What body structures and behaviors do students observe? What body structures

enable water snails to move? to avoid predators? Why do you think snails can go almost upside down without falling? Where do the snails spend the most time? How do they interact with their habitat? You may want to have students compare snails and Tubifex worms.

6. For older students (fifth and sixth grade), explain that the behaviors and body structures that help an animal survive in a habitat are called *adaptations.* Ask, "What are some human adaptations that help us to survive?" [the structure of our hands; our excellent sense of sight; speech; working together, etc.] Review some of the adaptations of snails and Tubifex worms.

7. Show students how to gently slide the snails off the surface of the cup and onto the surface of the water habitat. Students may be surprised to see that some of the snails float on the surface, inching their way towards the plants or toward the sides of the tank.

Snails have a strong muscular foot that clings tightly to the surface they move across. To avoid injuring the snail, slide the snail gently along the surface to release the foot's suction before lifting it free.

8. Have students take out their journals, or provide a new Aquatic Habitats student sheet. Ask students to draw and label the animals in their habitats, and write down their observations and predictions.

9. Collect empty cups, scratch paper, and wipe up any spills.

Going Further

1. Have students estimate how many worms are in the sand and in the gravel each day. They can keep a record of the distribution in their journals, and then observe what happens over time.

2. Have students observe a snail's movement—during a time when it is active—for a sustained period of time, such as one minute. How far did the snail travel? How did students determine that distance? If the snail were to continue at that pace, what distance would the snail cover in an hour? 12 hours? a day? How could they do an actual calculation of the snail's movement?

3. If you have microscopes, have students observe a Tubifex worm's body segments, bristles, digestive tract, blood vessels, and mouth.

4. For fifth or sixth graders: Have students make a list of various animals or plants. For each animal (or plant), have them do research to learn about adaptations and their benefits. Have students make a list of adaptations (either body structures or behaviors) in one column, and the benefit the adaptation provides in another column. This would make an excellent entry in student journals. For example:

	Adaptation	**Benefit**
bees	pollen baskets on legs	carry pollen to hive for food
	stinger	keep enemies away from hive
	long mouth part	sip nectar from flowers
	cooperation in hive	raise the young
wolf	warm-blooded	able to live in wide temperature range
	eyes in front of skull	depth perception to focus on prey
	sharp teeth	for tearing muscle
	live in packs	for protection; easier to capture prey

Date_____ Name The Tadpoles

Aquatic Habitats

Observations:

The worms are going under ground.
Some worms are wrapped around the plant.
The snails moved a 'zain.

Predictions:

The worms will make tunnles. The snails
will move more. The snails will die. The
worms will eat the snails. The plant
will move.

Activity 3: Fish Enter the Habitat

Overview

The hardy mosquito fish, or Gambusia (gam-BOO-zee-ah) has been introduced to ponds and wetlands throughout the world because it eats mosquito larvae. An adult fish can consume its own weight in mosquito larvae every day! The fish also eat a great variety of pond organisms, including Tubifex worms.

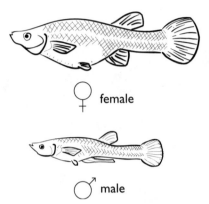

female

male

Please note: With students, we recommend you use the name "Gambusia," rather than "mosquito fish." This way you will avoid giving away the exciting fact that these fish eat mosquito larvae, and students can discover it firsthand in Activity 4.

In Activity 3, pairs of students carefully observe the structures and behaviors of fish in small, clear cups, before adding the fish to the model habitats. Older students reflect on how certain adaptations may help the fish survive. They predict what might happen when the fish are released into their tanks. Students will be riveted by the fascinating behaviors of the fish. Great excitement is usually generated as the fish eat the Tubifex worms!—and as some worms evade capture! This dramatic encounter with the food chain prepares students for Activity 4 when they explore the food web, the role of decomposers, and other interactions in their aquatic habitats.

As in previous activities, students make a detailed record of the habitat, drawing and labeling everything in their model ponds, noticing changes, and writing their observations and predictions. They then draw and describe the fish and discuss characteristics for distinguishing the individuals, as well as ways to compare their movements in the tank. They discover that *gills* are important body structures that allow the fish to obtain oxygen from water.

*We have provided an optional diagram of the body structures of a fish for students to label. While it may be tempting to introduce the parts of a fish right away, **it is best to wait until after the activity, when students have had an opportunity to discover many of these structures firsthand through their own investigations of real fish.***

What You Need

For the entire class:
- ❑ 3–5 gallon holding tank from previous activities
- ❑ bucket of dechlorinated water from previous activities
- ❑ about 3 dozen Gambusia fish (see "Getting Ready")
- ❑ one small aquarium net
- ❑ one container goldfish food (flakes). Any flake food will do. Goldfish food is usually the least expensive, and creates little odor.
- ❑ the Aquatic Habitats Questions on chart paper from previous activities
- ❑ newspaper and/or paper towels
- ❑ (optional) an extra aquatic habitat from previous activities to use in class demonstration

For each group of four students:
- ❑ aquatic habitat from previous activities

For each pair of students:
- ❑ a clear plastic cup containing 2 fish
- ❑ a piece of white scratch paper
- ❑ (optional) hand lens

For each student:
- ❑ a pencil
- ❑ a new Aquatic Habitats student sheet (master on page 22) or journal
- ❑ (optional) Parts of a Fish student sheet (master on page 43)

Getting Ready

Before the Day of the Activity

1. **Decide what kind of fish you will use and arrange for their purchase and delivery.** *Gambusia afinis*, or mosquito fish, are recommended for several reasons. Gambusia are extremely hardy, and do not require special aquarium equipment such as heaters or filters. Your students may even see them multiply—they give birth to live young, rather than laying eggs. They naturally prey on mosquito larvae, which students will study in Activity 4. In fact, they are frequently added to ponds throughout the United States by mosquito abatement agencies. Some of these agencies will provide teachers with Gambusia at no cost.

Gambusia are also available from some biological supply companies. (For more information, see "Some Sources for Key Items" on page 88.) Other hardy species that eat mosquito larvae are goldfish, guppies, and fathead minnows.

2. Try to schedule the arrival of your fish for a few days before they'll be distributed to students. Two days or so will allow the fish to recover from the stressful transport, and will improve their survival rates in the students' aquatic habitats. However, having all the fish in a crowded holding tank for more than a few days may result in some dying due to aggressive behavior by the larger fish. Also, be sure your holding tank is only about two-thirds full of water so fish will be less likely to jump out, and has Elodea sprigs to provide hiding places for smaller fish.

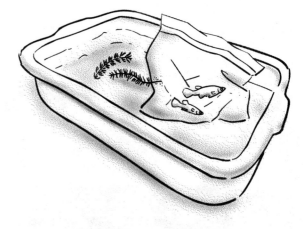

- You will need one fish per student, plus a few extra.

- When adding fish to the holding tank, be certain that the water temperature in the transfer container is about the same to avoid shocking the fish as you transfer them. One way to do this is to set the open bag or container of fish upright into the holding tank for about 10 minutes or so to equalize the water temperatures, before gently pouring the fish out of the bag.

- Feed the fish sparingly with flake food. Start with one or two flakes per fish. Feed them no more than they can eat in about 10 minutes. The fish may also nibble on the plants. Don't feed the fish immediately before the classroom activity. This way, the fish will be more likely to eat some of the Tubifex worms while students watch.

If you overfeed the fish, food will decay in the holding tank, and bacteria may multiply to the point where they start using much of the oxygen needed by the fish. Bacteria will also multiply if there are dead fish, so it is best to remove dead fish from the holding tank.

On the Day of the Activity

1. Half-fill a cup with dechlorinated water for each pair of students. Use the net to transfer two fish to each cup.

2. You may want to cover the cups with newspaper and set them aside to avoid students crowding around to see the fish.

3. Make a copy of the Aquatic Habitats student sheet (master on page 22) for each student or use folders or journals prepared earlier.

4. If you have decided to use them, make copies of the Parts of a Fish student sheet (master on page 43) for students to label after the activity.

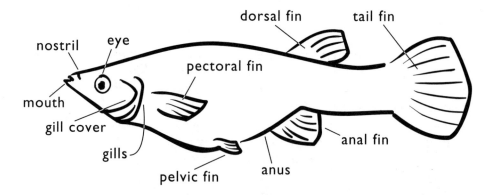

5. Have a sheet of white scratch paper and, if you've decided to use them, a hand lens ready for each pair of students.

GO!

Observing Fish

1. If you collected the tanks after the previous activity, distribute them to the teams of students.

2. Write "Gambusia" on the board and explain that this is the name of the small fish they are about to observe and add to their habitats. Help students with the pronunciation (gam-BOO-zee-ah). Tell students that each pair will get a cup with two Gambusia in it to observe. Make clear that **they should not put their fish into their aquatic habitats until you tell them it's time.**

3. Give the following tips for observation:

- They should put the cup on their desk or table so both students can easily see, with a piece of white scratch paper under the cup to make it easier to see the structures of the fish.

- So as not to frighten the fish, students should observe quietly and keep their hands out of the cups. In this way, they will be acting like scientists.

- Ask students to look at the parts of the fish, colors, and behaviors and quietly talk about how to tell the individual fish apart.

4. Give each pair of students a cup containing two fish. Also pass out the scratch paper.

5. Circulate among students, encouraging them to observe the body structures and behaviors of the fish.

6. After students have had plenty of time to observe and discuss their fish, regain the attention of the whole class and ask what they observed.

Fish Into the Habitat

1. Ask for predictions about what will happen when the fish are added to the tank.

2. Use an empty cup to model how students should gently lower the cup at an angle into the water so the fish can swim out into the habitat.

3. Ask students to observe the fish quietly without disturbing them. Say that one pair will release its fish while everyone on the team observes. After a few minutes, the second pair may release its fish.

4. Allow as much time as possible for observation of the fish in the model habitats, as well as observation of any changes in the habitats.

5. Have students take out their journals (or pass out new Aquatic Habitats student sheets), and allow time for students to draw their habitats with the fish, and to write down their observations and predictions.

6. Collect empty cups, scratch paper, and wipe up any spills.

Sharing Observations

1. Once again, regain the attention of the class. Invite several volunteers to tell what happened when they released the fish into their habitats.

2. Ask about other things they may have noticed in their model habitats. Where were the Tubifex worms? Are there any clues that the snails are eating something? Record their observations on the chalkboard.

3. When students mention certain body structures or behaviors, ask how they may help the animals survive in an aquatic habitat. [shell protects snail; fins help fish move; eyes on side of head may help fish watch for danger]

4. Help students understand that each plant or animal has different structures that serve different functions in growth, survival, and reproduction. For older students (fifth and sixth grades), emphasize the concept of adaptation—the idea that certain behaviors or body structures can be key to survival.

Predictions and Questions

1. Ask for predictions. What might change in the habitats in the next few days?

2. If students have questions, list them on the chart paper.

3. If no one mentions it, ask, "How do fish breathe?" Have students first focus on how we breathe by asking everyone to stand, and take a big breath. Ask students to notice how it feels as the air is pulled into their lungs. Have them slowly exhale, feeling the breath rush past their lips.

4. When students are seated, explain that our lungs take oxygen that we need from the air we breathe in. Say fish need oxygen too, but they do not breathe air with lungs. Instead, they make water flow over body structures called *gills*. The gills take in oxygen from the water.

5. Ask where students think the gills might be located. [on the sides, near the fins]

6. Sketch and—with student help—label the parts of a fish on the chalkboard or overhead. If you've decided to use them, distribute the Parts of a Fish student sheet for students to label or have students draw and label their own.

7. Encourage students to record more observations and drawings on the Aquatic Habitats student sheets or in their journals.

If students ask if their fish need flake food in addition to their worm diet, say that each team may add one or two flakes per fish once a day, under your supervision. Overfeeding is a big temptation for students. Emphasize that overfeeding can kill the fish. Always keep the fish food container in your possession.

Going Further

See the Outdoor Biology Instructional Strategies (OBIS) module entitled Water Breathers for a more detailed description of this activity. Related OBIS activities are referenced in the "Resources" section.

1. Use a medicine dropper and blue food dye to demonstrate the flow of water in and out of a tadpole and a crayfish in the following way:

a. Place each animal in a light-colored basin containing just enough water to slightly cover the animals.

b. When the animal has stopped moving around, slowly release a drop or two of dye near the area where the water is sucked into the animal's body. (For the tadpole this is the mouth. For the crayfish, this is the region where the shell covering the back joins the legs at the underside of the crayfish).

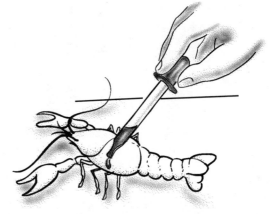

c. Don't tell students in advance where the dye will enter and exit the animals, but let them predict the currents.

d. The drawings illustrate how the dye comes out of a left gill opening on the tadpole, and out of the mouth area of the crayfish. (Dragonfly larvae also work very well for this.)

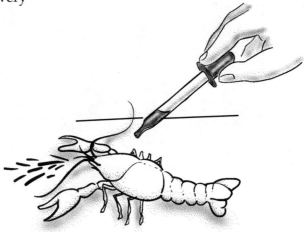

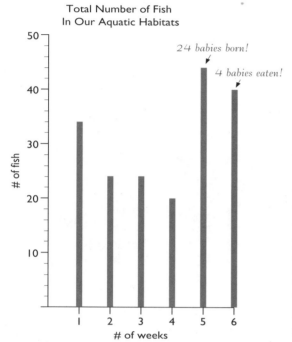

Total Number of Fish
In Our Aquatic Habitats

24 babies born!

4 babies eaten!

# of fish

# of weeks

2. Students are fascinated by the external and internal structures of fish. Smelt and other small fish, sometimes available fresh from a fish market, make excellent subjects for a fish anatomy activity followed by fish prints and even a fish fry.

3. Monitor the fish population by creating a graph for the total number of fish in the habitats. On the first day that the fish are introduced to the tank, record the total number of fish in all the habitats. Then, on a regular basis, such as weekly, repeat the count and record the number on the same graph. Did the population increase or decrease? If there are changes in the number of fish, what might be the causes?

4. For fourth through sixth grade students: Have students use their estimation skills to approximate the length of their fish using a common unit of measure, such as centimeters. Provide rulers or tapes to help with the estimation. Have students record all their estimates on the board. Have students identify the lowest and highest estimates (the range). Look for the most frequently occurring size estimate (the mode). Based on these estimates, have students determine the "average" size.

5. Have students write a story from the point of view of a fish (or other organism) in their habitat. Have them include things the fish does that might help it survive (explore, hide in shelter or keep near the bottom, move quickly, hold still, find food, etc.).

Parts of a Fish

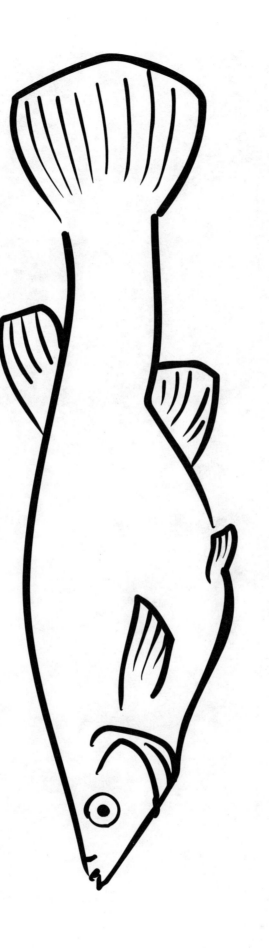

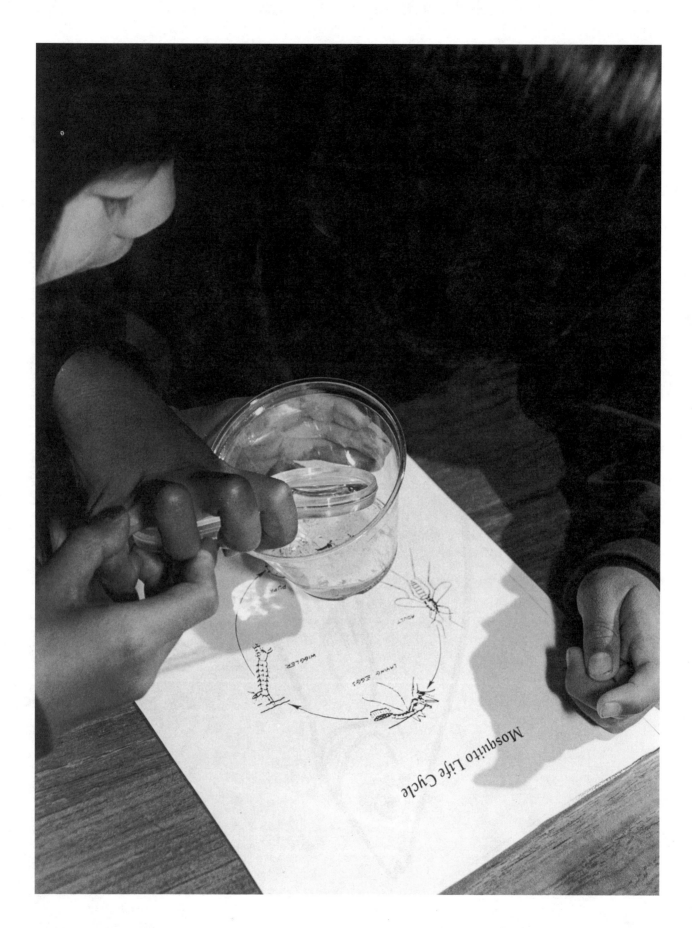

Activity 4: Mosquitoes

Overview

As their model habitats begin to resemble real ponds, your students begin to understand the complex interactions and balances of an ecosystem. They may have observed organic waste or dead animals beginning to decompose on the bottom, populations of worms or snails or Elodea expanding, shrinking, or dying out. Green "fuzz" may be visible on the sides or bottom, and the water may be turning green as the algae population increases. Students are ready to absorb the important biological concepts introduced more formally in this activity: the role of decomposers, life cycles, food webs, and biological control.

This activity introduces your students to the fascinating and infamous mosquito, which begins its life as an aquatic insect. Students compare the insect's life stages and structures, and observe the predatory behavior of the Gambusia as they snap up the larvae. Some of the larvae are kept separately in a class mosquito holding tank so they can develop into adults. Over the course of several days, your students may observe mosquito metamorphosis.

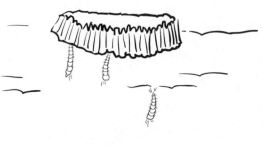

life size

A story entitled "Too Many Mosquitoes!" is provided following this activity as an optional class reading or homework assignment. If you're reading the story in the classroom, have students form a story circle or arrange themselves in a way that facilitates listening skills. Say that you will be reading a fictional story about mosquitoes and people. Suggest that students try to pick out clues from the story that could help solve the mosquito problems in their community. As you read the story, stop and include student comments and questions whenever you think them appropriate.

egg rafts & larvae

Following the story, invite questions and comments from the students. What examples from the story have students observed in their aquatic habitats? What is meant by the term "biological control"? What are some possible organisms that could be used as biological controls for mosquitoes?

If you need to break Activity 4 into two class sessions, we suggest you pause after adding mosquito larvae to the aquatic habitats.

Explain how you would like students to compose their own endings for the story. Give them guidelines on the nature and length of the assignment and encourage them to include information and description based on their observations of the aquatic habitats. For younger students, the assignment might be a "quick write" that draws from their recent experiences. Fifth and sixth grade students might research the biological control issues for their community.

What You Need

For the entire class:
- ❑ bucket of dechlorinated water from previous activities
- ❑ about 50–100 mosquito eggs or larvae (see "Getting Ready")
- ❑ a wide-mouthed container for a mosquito holding tank (a large plastic peanut butter jar would work well, as would a canning jar or a large plastic container—such as those used for bulk food items in large discount stores)
- ❑ a 12" tube of pantyhose (see "Getting Ready")
- ❑ a large rubber band
- ❑ a turkey baster or large syringe
- ❑ the Aquatic Habitats Questions on chart paper from previous activities
- ❑ newspaper and/or paper towels
- ❑ (optional) an extra aquatic habitat from previous activities to use in class demonstration

For each group of four students:
- ❑ aquatic habitat from previous activities

For each student:
- ❑ a clear plastic cup with a little dechlorinated water for 1–2 mosquito larvae
- ❑ 1 copy of the Mosquito Life Cycle student sheet (master on page 56)
- ❑ a piece of white scratch paper
- ❑ a pencil
- ❑ a new Aquatic Habitats student sheet (master on page 22) or journal
- ❑ (optional) hand lens
- ❑ (optional) 1 copy of the "Too Many Mosquitoes!" homework or class reading assignment (master on pages 57–59)

Getting Ready

Before the Day of the Activity

1. Several weeks in advance, order mosquito eggs or larvae from a biological supply company or create your own mosquito culture. If the weather is warm (above 40° F at night and 70° F during the day) and you live in a region where mosquitoes occur, you can create conditions to attract mosquitoes to lay their eggs. (See page 84.)

2. If you are ordering mosquito eggs, have them arrive at least 10 days prior to the activities so that the larvae have time to grow to an easily visible size. If you are ordering larvae, arrange for them to arrive a day prior to the activity.

3. For your extra mosquito larvae create a separate holding tank that will provide aeration and allow you to extract larvae while preventing the escape of any adults that develop:

   a. Fill a wide-mouthed container with dechlorinated water. Add some "green water," if you have some in the holding tank or the students' tanks (the larvae eat algae).

   b. Cut about a 12" length from the leg part of a pair of pantyhose, so that you have a tube open at both ends.

   c. Use a rubber band to attach one of the open ends of the pantyhose tube to the top of the container. The other open end is folded over the top to temporarily close the opening.

   d. When you need to take out some larvae, slide the baster through the opening, keeping the tube gathered around the baster.

4. Read the story "Too Many Mosquitoes!" to determine whether and how you will use it with your class.

Please see "Some Sources for Key Items." An "egg raft" contains about 100 eggs and costs about $10 from a commercial supplier. Ordering 100 larvae may cost as much as $40. As mentioned, you can create your own mosquito culture. If you have contacted a local abatement agency for Gambusia, you may be able to obtain mosquito eggs/larvae from them at no or minimal cost.

Read the "Behind the Scenes" section for a more complete description of the mosquito life cycle; also see the diagram on page 56.

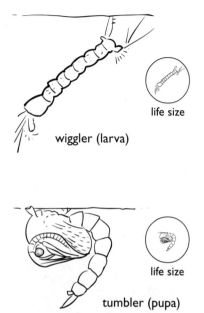

wiggler (larva)

life size

tumbler (pupa)

life size

Larvae (LAR-vee) is the Latin plural; larva is singular. Similarly, pupae (PU-pee) is plural, pupa is singular.

On the Day of the Activity

1. It's fine if all your mosquitoes are in the larval stage, but it's ideal if you have both larvae and pupae. Check to see what you have:

- **larvae ("wigglers")** Tiny larvae that emerge from eggs are commonly called "wigglers" because of the motion they make as they swim. Wigglers hang downward from or lie flat below the surface film of the water and breathe by means of gills and an air tube near the rear of their body.

- **pupae ("tumblers")** After several days, the larvae metamorphose into pupae that look like large commas resting at the surface of the water. If disturbed, they tumble away using their flap-like tails. They remain as pupae for only a few days before becoming adult mosquitoes.

2. Prepare a cup of mosquito larvae (and pupae if you have them) for each student:

a. Use a turkey baster to transfer 1–2 larvae and water to fill each cup one-quarter full.

b. If you have enough larvae, give each student 2 larvae that are different sizes (at a different growth point in their development).

c. If you have only a few pupae, try to make certain that at least one member of each group gets a pupa.

Note: Although each student will receive a cup of larvae, only two students per group will pour their larvae into their model habitat. The two remaining cups of larvae for each group will be returned to the class mosquito holding tank. This way, even if all the larvae in the students' tanks are eaten by fish, students may get to see adult mosquitoes develop.

3. Make a copy of the Aquatic Habitats student sheet (master on page 22) for each student or use folders or journals prepared earlier. Also duplicate a copy of the Mosquito Life Cycle student sheet (master on page 56) for each student.

4. Have a sheet of white scratch paper and, if you've decided to use them, a hand lens ready for each student.

Sharing Observations of Changing Habitats

1. If you collected the tanks after the previous activity, distribute them to the teams of students.

2. Begin with an invitation for volunteers to share observations about their changing habitats, first in small groups, and then as a whole class. Have them refer to their previous student sheets or journals.

3. Ask if student groups have noticed anything that has floated to the bottom of their habitats (dead leaves, fish or snail feces, dead animals). Ask what they think happens to this waste material. [the Tubifex worms eat it]

4. Point out that decaying waste is very important in aquatic (and other) habitats. The Tubifex worms—and bacteria that are too tiny to see—help *decompose*, or break down the decaying waste into nutrients for plants to use.

You may want students to use the scientific term detritus (dee-TRY-tus) for the organic waste in their habitats.

5. Ask if anyone has noticed their water turning green, or green "fuzz" growing on the sides or bottom of their tanks. Explain that this is a green plant called *algae.* Ask, "What do all green plants need to grow?" [sunlight, air, and nutrients] Emphasize that their Elodea and algae need sunlight, air, and the nutrients made available by decomposers breaking down decaying matter.

6. Review the class list of Aquatic Habitat Questions. Have any of the questions been answered? Are there any new questions to add?

Observing Mosquito Larvae and Pupae in Cups

1. Ask students to raise their hands if they have ever been bitten by a mosquito. What do students know about mosquitoes? Ask, "Are there mosquitoes around here?"

2. Explain that before they become adults, young mosquitoes live in aquatic habitats, and are found in almost every fresh water habitat in the world.

3. Tell the class that each student will get a cup with one or two young mosquitoes to observe and sketch. Ask students to compare their insects with others in their group. **Tell students not to add them to their habitats until you tell them to do so.**

4. Give each student a cup of larvae and pupae, and a piece of white scratch paper to put under the cup to enhance visibility.

5. After students have had time to observe, regain the attention of the class. Ask students to describe and compare their larvae.

6. As students share observations about how the larvae and pupae move, ask, "How do you think that might help them survive?" [it might help them escape getting eaten]

7. Explain that the ones that wiggle are called "wigglers," or larvae. The ones that tumble or somersault are called "tumblers," or pupae. Write these words on the chalkboard and explain that the singular form of larvae is larva and of pupae is pupa.

8. Briefly explain the life cycle: Female mosquitoes lay eggs that hatch into larvae. The larvae change into pupae, the pupae change into adult mosquitoes. These adult mosquitoes in turn lay eggs, and the cycle repeats.

Adding Mosquito Larvae and Pupae to Aquatic Habitats

Although the fish will probably eat all the mosquito larvae, some may survive to adulthood. Students should keep the lids on their tanks; if your lids have the small pinholes, the adult mosquitoes won't escape.

1. Ask students for predictions about what might happen when the larvae and pupae are added to their model habitats.

2. Say that each group will **pour two cups in the tank, but keep the rest of the larvae and pupae** to watch them turn into adult mosquitoes.

3. Review and model the procedure:

   a. Each team will decide together which two students' larvae/pupae will be poured in and which two will not.

   b. Two students will gently tip the larvae from their cups into the tank.

   c. Everyone in the group will observe what happens.

   d. The other two students in the team will return their cups of larvae back to the area of the class mosquito holding tank.

4. Remind teams to put the lids on their tanks when they're finished adding their larvae, and point out that the small holes will trap any flying mosquito adults that may develop, but allow air into the container.

5. Circulate among the groups. You may need to remind them to behave like scientists and use quiet observation techniques.

6. Have students take out their journals (or pass out new Aquatic Habitats student sheets), and allow time for students to draw their habitats with mosquito larvae or pupae, and to write down their observations and predictions. Also have students record some of the events they observed, the number of larvae added to the habitat, descriptions of fish interacting with the larvae, wiggler movements, etc.

7. Let students know that while they're working you'll be distributing a copy of the mosquito life cycle to each of them. They can add the sheet to their journals and use it as a reference. Pass out the Mosquito Life Cycle student sheets.

8. Transfer the "saved" larvae from the cups into the holding tank, using a syringe or turkey baster. Point out to students that you fold over the covering when you're finished adding the larvae, because the small holes in the pantyhose will trap any flying mosquito adults.

9. Collect extra cups, and wipe up any spills.

10. As possible, provide time in the next few days (or weeks) for continued observations of and predictions about the aquatic habitats and the mosquito holding tank, with continued journal writing and record-keeping.

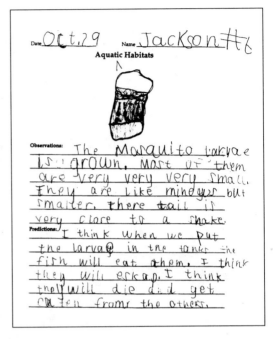

If you have decided to make Activity 4 into two class sessions, this is a good place to break.

Discussing Observations and Introducing Food Webs

1. Gather the class away from their habitats for the discussion. Invite several students to share observations.

2. Reveal the other name for Gambusia—mosquito fish! Say that in some places county or state employees add mosquito fish to lakes and ponds to control the population of mosquitoes. This way they don't have to use poisons.

Ask why we might not want to add poisons to aquatic habitats. [other animals could die] You may want to introduce the term *biological control* and explain that that's the name given to any organism used to control the population of another organism.

3. Ask students what each organism in their tank eats (besides the fish food they may be adding). As they answer, sketch the organisms on the chalkboard and **draw arrows to show the direction the nutrients go.** Here are some examples: (➔ = eaten by)

mosquito larva ➔ fish

Tubifex worm ➔ fish

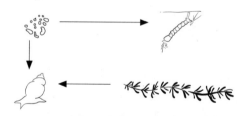

algae ➔ mosquito larva

algae ➔ snail

Elodea plant ➔ snail

As you add organisms, the arrows may begin to form a "web" on the chalkboard. Help students appreciate the complexity of life in even this small system. For middle school students familiar with the concept of energy transfer, introduce the idea that energy is transferred along the food chain.

4. Ask, "What eats the fish in your habitats?" [nothing eats the live fish] Explain that, if fish die, decomposers like Tubifex worms and tiny bacteria break them down to release nutrients for plants. Also, when their feces or waste falls to the bottom, decomposers break this waste down into nutrients too. Add an ➔ from a fish to the Tubifex worms.

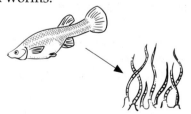

5. Ask, "What 'eats' the snails?" [It works the same way as the fish.] Draw snails → worms.

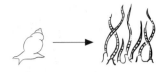

6. Have older students help you draw an example of a nutrient cycle:

fish → worms → bacteria → algae → mosquito larva → fish →

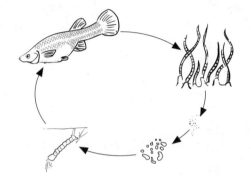

or

snails → worms → bacteria → Elodea → snails →

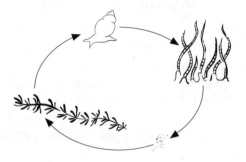

7. With older students, extend the food web beyond the classroom habitats by showing in a drawing how the Gambusia could be eaten by bigger fish, which could be eaten, for example, by people, or by bears. Or mosquitoes could feed on people!

8. To emphasize that algae, Elodea, and all plants need sunlight, as well as nutrients and air, to live, draw the sun shining on the chalkboard food web.

Although understanding what eats what and learning how to create nutrient cycles are important, the larger goal is for students to begin to appreciate the complex interactions among organisms and their habitats that make up an ecosystem.

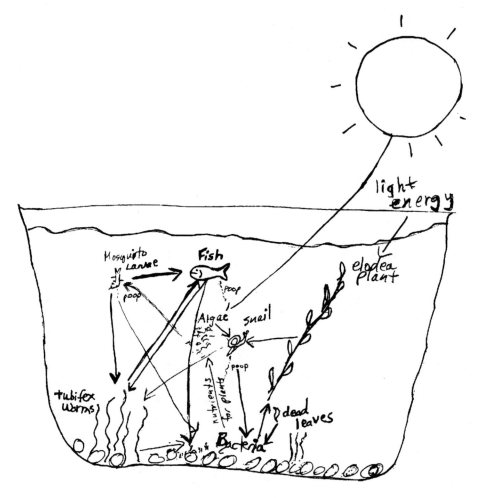

Labels within the illustration:
light energy
Mosquito Larvae
Fish
poop
poop
Algae
snail
elodea plant
poop
tubifex worms
nutrients for plants
dead leaves
Bacteria

Sample Food Web in the Aquatic Habitat

9. You may want to have students draw some food webs. Depending on your students' grade level, you may encourage them to sequence just two organisms at a time, as in #3, above, or create whole cycles or webs, including decomposers. Encourage students to create drawings of food webs in other habitats (for example, bacteria, earthworms, grains, mouse, owl).

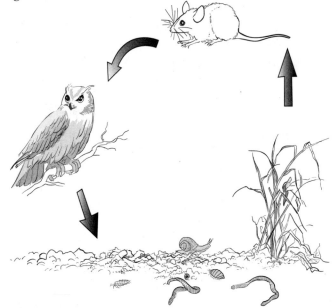

Going Further

1. Encourage students to explore the interdependence of the organisms in other habitats by reading and discussing the children's ecological tale, *The Day They Parachuted Cats On Borneo* by Charlotte Pomerantz or *The Old Ladies Who Liked Cats* by Carol Greene (both of which are annotated in the "Literature Connections" section).

2. For fifth and sixth graders, have students assume the roles of mosquito abatement scientists hired by a fictitious Chamber of Commerce because flooding has led to a dramatic increase in mosquitoes. Chemical sprays must be avoided because of their possible dangerous side effects to humans and the environment. Each student team should research the pros and cons of various mosquito abatement methods for their community and report back to the class. Some possible methods include: dissemination of mosquito fish to local residents with ponds; spraying wetlands with a chemical that prevents the mosquitoes from growing up; spraying wetlands with naturally-occurring bacteria that attack the mosquito eggs; spraying wetlands with nematode (tiny roundworm) eggs that will hatch and eat the mosquito larvae; draining roadside ditches; neighborhood clean up of containers that collect water.

3. Use an overhead projector to compare the movements of the fish, worms, snails, and mosquito larvae. Place several clear deli-type containers—each containing an organism and a small amount of water—on the surface of the overhead. Turn on the lamp and observe the animals.

4. Have students research the major kinds of mosquitoes and the methods of preventing the spread of diseases such as malaria, dengue, yellow fever, and encephalitis. The story of the conquest of malaria by researchers is one of the most important chapters in the history of science and medicine.

5. The introduction of mosquito fish to Australia and areas of North America has resulted in the decline of native fish and insects. Have students research places where biological control efforts led to unforeseen problems, such as cane toads in Australia and the mongoose in Hawaii.

Mosquito Life Cycle

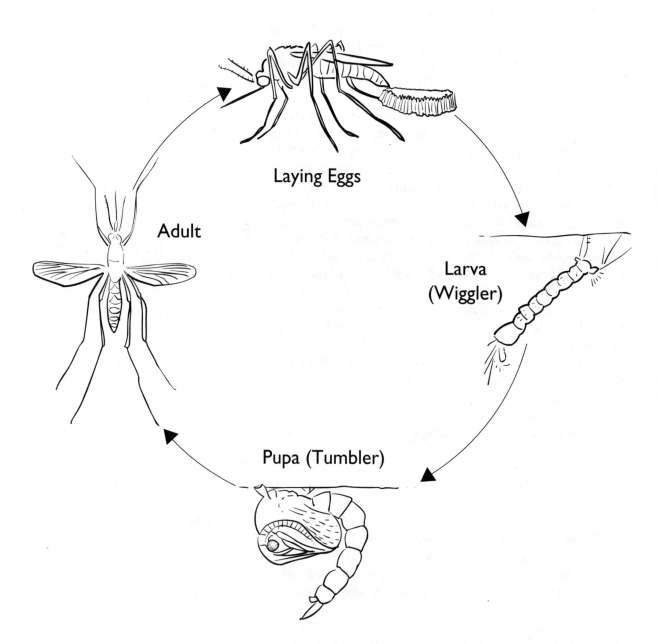

Laying Eggs

Adult

Larva
(Wiggler)

Pupa (Tumbler)

"Too Many Mosquitoes!"

Life has never been easy for mosquitoes, and it certainly hasn't been easy for their neighbors. If you've ever been bitten by a mosquito, you know how those bites can itch! This story begins millions of years ago when mosquitoes spent most of their lives wiggling around the ponds and lakes that covered the land. The sunlight made the water warm and provided light energy for the algae and water plants which the wigglers would eat.

But the mosquito wigglers had a problem. Everywhere they wiggled, hungry fish followed and snapped them up. A few wigglers always escaped by holding still like sticks and by hiding under leaves. Their underwater neighborhood was very dangerous and the wigglers who grew up fast, were the ones who could fly away and stay alive. The speediest ones grew from egg to adult in less than two weeks!

Even the wigglers that made it to the top of the pond to fly away on their new mosquito wings had to watch out for dragonflies and frogs with sticky tongues that could snatch them as they flew by. Long ago, female mosquitoes discovered something to eat in the meadows and woods away from the water predators. The mice and rabbits had tasty warm blood that provided the female mosquitoes with a great high-protein meal for their developing eggs. In addition to sipping plant juices the way their male mates did, the hungry females used their sharp straw-like mouths to suck their victim's blood. "Yum!" they hummed.

"Ouch! These mosquitoes make us itchy," complained the mice and rabbits! Happily for the animals, soon great flocks of birds zoomed back and forth across the sky catching the mosquitoes in their beaks. Around the world, the swifts and swallows and martin birds became known as champion mosquito eaters.

"Zzzeee—Ahoy all mosquitoes, Zzzee." Male and female mosquitoes from miles around were attracted by this high pitched call to action. "We need a new plan, these swift

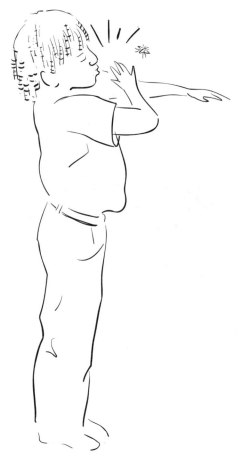

birds are too fast and hungry." Some mosquitoes began hiding during the day, then flying at night when it was easy to bite people and animals while they slept.

"Ouch, slap, slap! This is terrible!" complained the people. "We are covered with itches!" In some places the clouds of mosquitoes were so thick at night that they flew into peoples' mouths and eyes! How would you feel if you lived where mosquitoes were so thick you could hardly breathe?

As the billions of mosquitoes bit them night after night, millions of people and animals got sick from diseases. You see the mosquitoes didn't wash their sucking straws in between meals, and sometimes germs that cause diseases like malaria were passed on.

All across the Earth, groups of people got together to figure out what to do about the mosquitoes. Some people found that body paint and perfumes kept the mosquitoes away. Some villages moved away from the ponds and lakes to places that were high and windy. Some people invented woven screens called mosquito nets to keep mosquitoes out of their houses, and bats and frogs were welcomed as mosquito eaters of the night.

Thousands of years passed and still the battle between mosquitoes and people and animals raged on. People kept thinking of new ways to kill the pesky insects. For example some people turned to the animals for help, keeping small lizards in their homes to eat the mosquitoes, and building beautiful bird houses for mosquito-eating birds like swifts and martins.

Then about a hundred years ago, people figured out that mosquitoes come from the little wigglers in the water. Someone discovered that pouring oil across the ponds and marshes killed the baby wigglers. The oil plugged their breathing tubes and they couldn't breathe. That worked pretty well, and there were fewer mosquitoes, but all of the fish were killed too.

Unfortunately for the people, some wily mosquitoes always got away and laid more eggs. Without the hungry fish to eat the wigglers, ten times more mosquitoes flew from their water homes to snack on their warm-blooded neighbors.

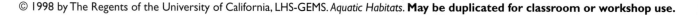

"We'll fix them!" said the people. "We'll poison them off the face of the earth!" And for years and years the people sprayed poison across the ponds, marshes, and meadows of the world. They tried many different kinds of poisons, but each time some hardy mosquitoes survived and laid thousands of eggs. Within a very short time millions of mosquitoes were back sucking their blood meals.

The mosquitoes were everywhere! Many weren't affected by the poisons and there were fewer and fewer animals to eat them. Mosquitoes hummed across the land. Millions of people were still getting itchy and sick with diseases, while there were fewer animals that ate mosquitoes in many areas. Some kinds of animals were wiped out by the poisons and many more lost their homes when marshes were drained.

The struggle against the mosquitoes seemed almost hopeless, but the people didn't give up. They asked scientists for help. The scientists learned as much as they could about the mysterious lives of the mosquitoes by finding the answers to questions like: Where do they lay their eggs? What diseases make them sick? How do they change into adults? What eats mosquitoes?

The scientists developed a strategy called "biological control." They started searching for the natural predators of mosquitoes and helped increase the numbers of those predators. You can help too. You have been studying the mosquito and have some ideas for solving the problem. The story is now in your hands. Please write or draw a successful ending to the problem of "too many mosquitoes."

Activity 5: Further Explorations

Overview

This activity offers ideas for further investigations in the classroom and on a pond field trip, with many suggested ways to deepen your students' understanding of habitat, food web, adaptations, and ecosystems. Choose what works best for you, or do it all!

We've divided these further explorations into two main sections—Part I, ongoing investigations with the aquatic habitats **in the classroom** and Part II, suggestions for activities to be conducted on a **field trip** to a nearby pond, lake, stream, or other natural aquatic habitat.

Part I includes suggestions for students to design and conduct experiments with the aquatic habitats, make closer studies of the mosquito life cycle and/or decomposition, and observe changes as the habitats are dismantled.

Part II includes suggestions for helping your students take in the beauty of the habitat and focus on careful observation. Sampling activities are suggested and students are encouraged to connect their field observations to their classroom habitat upon their return. Other specific possible activities suggested are: crayfishing, can fishing, and studying water striders.

Part I: Ongoing Investigations with Classroom Aquatic Habitats

The classroom habitats make a wonderful setting for ongoing student investigations—they serve, as they have throughout the unit—as a living model of similar habitats in the real world. Scientists use models frequently for careful study of something that may be hard to observe in nature. Whether or not a field trip to a real pond is in your plans, your students will learn a lot by using their model ponds for further investigations in the classroom that spring from their own questions and speculations. Following are some ideas to help get these investigations started.

Recording New Developments

1. Over a period of several weeks have students document observations with journal entries and drawings, perhaps including:

- appearance of green algae

- growth of Elodea

- transformation from larval to pupal to adult mosquitoes

- snail trails

- appearance of snail egg masses and young snails

- birth of young mosquito fish

- appearance of young Tubifex worms

- recycling of dead leaves, mosquitoes, and animal waste

- drop in the water level and rings of white residue above the water level

- change in the color of the water

2. Encourage students, as they make and record their observations, to be thinking about questions they have and further investigations they may want to make.

Designing and Conducting Experiments

1. Based on their own observations and questions, and with your feedback and suggestions, ask students to come up with a question to investigate.

2. The following are some examples of interesting questions that students might investigate.

- How does adding another organism such as a bull-frog tadpole, a crayfish, or some Daphnia affect the habitat and the other organisms?

- Do the animals prefer light or dark areas of water?

- Are the fish attracted by bright colors? by movements?

- How might an increase in natural light influence life in the aquarium?

- How do the animals react to different substrates, such as sand and rocks?

- Which areas of water, (top, bottom, or middle) do the different organisms prefer?

3. Here is a suggestion for one way that a group of students can investigate the question: Do fish prefer certain areas of the habitat?

a. Have teams perform a two-minute, focused observation of a fish in the habitat.

b. Students could record the location and behavior of a particular fish every 15 seconds for two minutes. One student will be the timekeeper, one the recorder, and two can be observers.

c. Each time the timekeeper calls "time" the recorder should make an X on a map of the aquarium showing the location of the fish. Observers should write down what the fish is doing at that time.

d. Challenge students to distinguish between factual *observations* (the fish are swimming, hiding in the shelter, eating, moving fins, etc.) and *inferences* or assumptions that are not directly based on observation (the fish is frightened, hungry, etc.).

To extend this activity, you might want students to use or adapt the activities from the GEMS guide Mapping Fish Habitats. *The GEMS guide* Animals in Action *also includes emphasis on learning the distinction between observations and assumptions.*

4. Students may also wish to make a closer study of the mosquito life cycle. They could put mosquito larvae from the holding tank into individual covered cups for closer observation of the metamorphosis from larvae and pupae into adult mosquitoes and make a detailed record of each stage.

Decomposition

1. If a fish has died, some students may be interested in observing the process of decay and making a closer study of the process of decomposition.

2. To keep smells to a minimum, put the dead fish in a container with some aquarium water, then cover with a lid that has a few small perforations. Place in a well ventilated location.

3. Students may also want to experiment to see if Tubifex worms will gradually consume the body.

Observing Changes as Habitats are Dismantled

1. When you near the end of your schedule of aquatic studies, provide time for students to disassemble the aquariums. (See page 80 for suggestions on how to dispose of living organisms.)

2. As the groups work together to return the fish, snails, worms, and Elodea to the classroom holding tank, encourage everyone to look for changes in their tanks.

The following questions may help focus student observations:

- What changes do you notice in the animals?

- Compare the aquarium water to tap water. How is it different?

- What changes do you notice in the bottom gravel and sand, the shelter and the walls of the aquarium?

- Has your model habitat become more like a real pond? If so, in what ways?

- How does your model habitat compare with your predictions?

3. Have students write a closing essay or summation in their journals. This could summarize their observations throughout the unit and discuss what they think are the most important things they have learned. What did they learn by watching the changes over time? What conclusions can they draw about the ways organisms in an ecosystem interact with each other? What surprised them most? What questions would they most like to further investigate?

Part II: An Aquatic Field Trip

A field trip will give students an exciting opportunity to apply their knowledge while exploring a diverse natural environment. Following our initial suggestions for the field trip, we provide detailed information on three "Going Further" activities for you and your students to consider including on your field trip: Crayfishing, Can Fishing, and Water Striders.

The water's edge is a habitat rich in species diversity. Both aquatic and terrestrial organisms may be found in the foot-wide border around ponds, lakes, streams, marshes, and other wetlands. During the warm months of the year, organisms such as frogs, tadpoles, fish, water bugs, diving beetles, snails, dragonflies, and water plants abound in wetland habitats. Hidden beneath the water, and not as well known, are the organisms that live in the sediment on the bottom. Tubifex worms, clams, crayfish, seed shrimp, flatworms, and insect larvae are some of the harmless organisms that students may find under rocks and in the mud.

What You Need

For the entire class:
❑ several copies of the Guide to Freshwater Life (master on pages 76–78)
❑ several light-colored dishpans or clear, plastic sweater boxes to use as observation trays for large organisms
❑ *(optional)* a small dry-erase (white) board and marker for recording class observations at the pond

For each group of four students:
❑ 4 clear plastic cups
❑ a dip net or a small (about 4" in diameter) metal strainer with a handle
❑ 2 disposable, white cereal bowls or other light-colored basins to use as observation trays
❑ *(optional)* 2 hand lenses or bug boxes

Getting Ready

Before the Day of the Field Trip

1. Visit the site yourself so you can better plan for excited students to do their investigations safely and without harming the habitat. Most parks and reserves have strict rules for protecting the plants and animals. Make certain you know the local regulations and are aware of any endangered species.

2. Arrange for several adult volunteers to come along. Make sure they are clear on their roles and know what is expected of students.

On the Day of the Field Trip

1. It is important to establish guidelines before the trip. Once at the pond, it is difficult for students to focus on anything but their investigations.

2. In the classroom, go over the following guidelines with your students to help them gain respect for the habitat:

- Try not to trample the plants; they provide the food for the system and prevent erosion. Stick to trails.

- Don't take any samples of plants around the pond. If possible, draw the plant or a leaf.

- Be careful not to harm organisms in the net by squishing them or keeping them out of water too long. Model for the students how to invert the net to remove the animal.

- After you check a rock or log for living things, replace it just as it was before so that thousands of small organisms aren't made homeless.

- Once they have been observed, return all animals to the places they were collected.

- Everyone will see more wildlife and enjoy themselves more if the group talks and moves about the habitat quietly.

3. Be sure to keep in mind the following safety issues, including all necessary precautions to be taken around water.

- Establish clear boundaries, and tell students that they must remain within sight and sound of their adult leader at all times.

- Use a buddy system for keeping track of class members.

- Be on the lookout for poison ivy, poison oak, and poison sumac. These plants love moist areas and just one careless student can inadvertently spread the plant oils to many of their classmates.

- If your field trip site is adjacent to large areas of woodland or tall grass interspersed with bushes, have students take precautions to avoid ticks such as wearing closed shoes, long pants, long-sleeved shirts, and light-colored clothing so ticks can easily be seen; tucking pant legs into boots or socks, and tucking shirts into pants; applying insect repellent containing permethrin to pants, socks, and shoes; and checking themselves frequently for ticks during and following the field trip.

- If there are poisonous snakes in your area, take appropriate precautions.

4. You may also want to discuss what they should wear and what to bring and not bring along on the trip.

One teacher suggests having two adults serve as the end boundaries, with others spacing themselves out in between students.

GO!

One teacher recommended the following guided observation:

• *Have students sit or stand at arm's length from each other, where they can observe the pond from a distance.*

• *Have them close their eyes and be totally silent for 15 seconds.*

• *With their eyes still closed have them listen for two or three sounds made by humans.*

• *Have them listen for four sounds made by nature.*

• *Listen for the highest sound, the lowest.*

• *Count the number of seconds of complete silence. How long was it?*

• *Have them open their eyes and look around. Is there anything here they weren't expecting to see? Any other observations?*

It is usually best to model the procedures away from the pond because it is difficult to hold the students' attention once they are near the water's edge. It is also best not to distribute the materials until it's time for students to begin the activity.

Sensing the Beauty of the Habitat

Begin the field activities *away* from the water's edge with a three to five-minute, whole-class sensory observation in which students sit or stand quietly and hear, smell, feel, and see as many sensations as possible, while reflecting on the personal emotions stimulated by the aquatic habitat. This quiet introduction to the environment will calm students and enable them to witness animal life that quickly hides from noisy intruders.

Introducing the Procedures

1. Away from the water's edge, review the boundaries, safety procedures, and other rules designed to minimize impact on the environment.

2. Let students know that they'll actually be exploring several parts of the aquatic habitats to find out what lives in them. Explain that, for example, the organisms that live in the center of the stream or pond may be different than those living at the edge of the water.

3. Tell students that there are some ways of sampling habitats that are easier than trying to net those organisms directly. They can, for example:

    a. Gently sweep the net through plants in the water, then empty the contents into an observation tray.

    b. Lift a medium-sized rock and place it into a tray so that the creatures clinging to the underside of the rock can be seen.

    c. Scoop up some bottom sediment and strain out the mud by repeatedly dipping the net in the water. Transfer any plants and animals to an observation tray filled with water.

4. Remind students to try all of the different habitats: the surface, the sides, the bottom, and the vegetation growing in the open water. When they have samples from two slightly different water habitats, they should compare what they see.

Sampling the Aquatic Habitats and Sharing Findings

1. Give each group a net or strainer, four plastic cups, and two observation trays. Challenge teams to find as many different kinds of plants and animals as they can. Point out that one of each kind is enough, and that they should share their findings with other groups.

2. Circulate among the student groups, encouraging team-work and sharing of discoveries. Allow as much time as possible for the sampling activity.

3. Gather students away from the water's edge. Ask students to arrange their observation trays in a large circle so that everyone can walk around them and look over the contents.

4. Gain the attention of the whole class and ask them to help you list the plants and animals that were found. If you have a small dry-erase (white) board, you may want to list their discoveries on it. You may need to help students come up with descriptive names for some of the organisms whose actual name is not known. That is fine! Students may be able to identify organisms with the help of the Guide to Freshwater Life (master on pages 76–78).

Allow as much sharing as possible, but keep the discussion moving and stop when the group begins to be distracted. More sharing can happen back in the classroom.

5. If time allows, have students resume their sampling with the following questions in mind:

- Which habitats had the most animals, and which had the most plants?

- Which types of organisms were caught most often?

- Which organisms were the hardest to find? Why?

6. Have the teams return their organisms to the water where they were found.

Making Connections Back in the Classroom

1. Back in the classroom, give students an opportunity to share their discoveries. Ask, "How was the pond like your habitat?" "How was it different?"

2. Ask, "What might happen if all the fish in the real pond died?" [Animals like mosquitoes and Tubifex worms

Scientists make use of models in many different ways. The classroom aquatic habitat has served you and your students as a model of an actual pond. Unlike some other models, such as those on a computer, or those that model the solar system, this classroom model has real life in it—the organisms grow and interact, although the conditions are not exactly the same as they are in nature. Models have many advantages and many limitations. Teachers of 5th and 6th grade students may want to extend this discussion and could consider presenting the GEMS guide River Cutters, *which explores issues raised by models in greater depth.*

You might encourage students to make a drawing or mural of a real pond.

You might encourage students to make a drawing or mural of a real pond.

As another journal entry or writing assignment, you could also have students choose one aquatic organism and write a fictional story from the point of view of that organism. A similar assignment in the GEMS guide Terrarium Habitats *is entitled "An Isopod's Journal" and has served as an excellent assessment of student learning in that unit. "The Journal of a Tubifex Worm" could do the same for this unit.*

might become more numerous; fish-eating birds might leave the area.] Help students understand and better appreciate the complexity and ecological balance of the pond food web.

3. Conclude by having students write about the aquatic field trip in their journals. Fifth and sixth grade students could list the adaptations of organisms they found and hypothesize their uses, as in Activity 2. All students could be asked to compare and contrast the real-world pond with their classroom models of a pond habitat.

Going Further with the Aquatic Field Trip Activities

The activities described here are a sample of the possible further investigations that you and your students may choose to undertake at the site. Each activity lists the appropriate equipment and procedures.

Crayfishing

Crayfish are widespread throughout the country. If you suspect they inhabit the planned field trip site, take along some bait and fishing line. Fishing for crayfish (or "crawdaddin' ") attracts a large following. People of all ages enjoy trying to catch these small freshwater relatives of the crab and lobster.

There are over 200 different kinds of crayfish in North America. You can find them in a variety of freshwater aquatic sites, from swamps to rivers. Crayfish seek cool water among rocks, snags, and other bottom objects, including human-caused debris. Crayfish are scavengers that feed on almost any kind of edible matter they encounter.

Choose a site about 50 yards long with gently sloping banks and plenty of elbow room. Avoid deep, swiftly flowing streams or rivers. The best sites usually have snags, rocks, or tree roots for crayfish to hide among. If possible, check out the site prior to your group's visit using the fishing technique described below.

What You Need

For the entire class:
- ❏ bacon, cut in 2" strips, makes good bait but you can also use pieces of meat
- ❏ *(optional)* pieces of old nylon stockings for holding soft bait other than bacon so that it can be attached to the fishing lines
- ❏ *(optional)* weights (e.g., washers) to sink the bait
- ❏ *(optional)* blue food dye and a medicine dropper to investigate water currents around the crayfish once they have been caught

For each group of two or three students:
- ❏ a 7–10 foot piece of string or fishing line
- ❏ a medium binder clip
- ❏ a dip net
- ❏ 1 dishpan or bucket
- ❏ *(optional)* a 3 foot long thin dowel or piece of bamboo to use as a fishing rod

Demonstrate Setting Up a Crayfishing Rig

1. Securely tie a binder clip to the end of the line. You can fish with just the line and clip or you can attach the line to a fishing pole.

2. Select a piece of bacon or other bait, and securely clamp it with the binder clip.

3. Tell the students that to catch a crayfish, they will drop or toss the bait into the water and let it sink to the bottom among rocks or tree roots.

4. When they see or feel a crayfish grab the bait, they should bring the bait and crayfish *slowly* to the surface, and then use a net to scoop up the animal while it is still hanging onto the bait.

5. Explain that if the students try to lift the animal out of the water without using the net, the crayfish will usually let go of the bait and escape.

*We recommend **against** students or teachers handling the crayfish—nets work well to transfer them. For your information, should it prove necessary to pick up a crayfish, reach from the rear and grab the top of the back (carapace) just behind the big front claws. In this position, the crayfish cannot reach back to pinch you. They have tricks to escape their raccoon predators, so you may be surprised by sudden flips of the tail!*

Many methods can be used to fish the cans out of the water. You could pick them up with your hands (wearing gloves for protection), use a long-handled net, use a strong magnet tied to a length of twine, or bend a wire coat hanger into a horseshoe shape to push and "clamp" onto the side of a can. This latter device is called a Can Grabber in OBIS activities.

Observing Crayfish

1. Once a crayfish has been netted and transferred to a basin of water, notice how it can quickly reverse direction and scoot backward. There may be small organisms clinging to the sides of the crayfish such as tiny leeches.

2. You may want to test for "breathing" currents around the crayfish by releasing small drops of food dye near the base of the crayfish's legs. The dye will be pulled into the gills and expelled near the mouth. Please also see the first "Going Further" on page 41 of Activity 3.

Can Fishing

Discovering the aquatic animals that live in discarded cans and bottles in your area can be a fascinating activity. These artifacts of the "litterbug" often become underwater homes for a variety of aquatic animals, such as minnows, crayfish, snails, scuds, and small catfish. Most animals are particularly vulnerable to predation during their early development. The narrow openings of cans and bottles can exclude potential predators and improve a young animal's chances for survival.

If your field trip site has some submerged cans, you and your students will enjoy discovering what kinds of aquatic organisms live in or on these accidental nurseries.

What You Need

For the entire class:
❑ work gloves for any adults or students who will handle the cans
❑ a long-handled net
❑ a light-colored basin for observing the "catch"
❑ *(optional)* a hand lens

Making and Observing the Catch

1. Challenge the students to catch as many cans as possible and to discover what kinds of aquatic organisms live on or in submerged cans. Encourage them to think about why organisms might make their home in a can.

2. To retrieve the cans, students should gently scoop the net under the can and pull it slowly out of the water, then place it into the observation tray.

3. Students should pour the contents of each captured can into the observation tray and then set the can itself in the tray to observe any organisms that are attached to the outside of the can. Keep can organisms moist and sheltered from direct sunlight.

4. After most of the students have "caught" one or more cans, call them all together and let them share what they have caught. Ask questions such as:

- What kinds of aquatic organisms did you find living in or on cans?

- Did some cans have no organisms? Why might that be?

- What advantages might a can offer as a home for certain organisms? what disadvantages? [If students don't suggest it, you may want to point out that many aquatic animals are particularly vulnerable to predation during their early development and that the narrow openings of cans and bottles can exclude potential predators and improve a young animal's chances of survival.]

5. Tell the students they now have a problem to solve. Now that they have discovered that certain organisms use submerged cans as homes, should they dispose of the cans as litter, or return the cans to the water because they house aquatic organisms? A good question to start the discussion is: "If the cans were removed, where might the organisms live?" Suggest that students search for organisms living in or on natural materials before making a decision. Depending on the discussion and your own preferences, allow individual students to make their own decision, or decide as a class.

6. After everyone has had a chance to observe the organisms, have the students return the contents of the observation trays to the water where they were found.

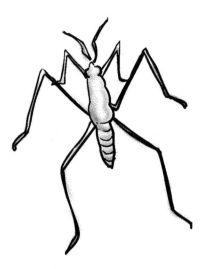

Water Striders

The darting movements of the water strider is a common sight at many streams and ponds. If your site has these remarkable insects skipping across the pools, your students will enjoy investigating their movements and feeding behaviors.

Striders are harmless to humans. They move across the water using their middle legs as oars, and steering with their rear legs. A strider can walk on the water because it has long legs covered with hundreds of tiny hairs that distribute its weight over a large area of the water. The surface tension of the water supports striders just as it can support a carefully placed sewing needle.

Water striders are voracious feeders, eating insects (dead or alive) and other tiny animals that land on the water's surface. Like all members of the waterbug family, striders have long, thin beaks for mouths. The beak is used as a straw to suck body juices from their prey. Striders locate their food by both sight and their ability to detect the vibrations tiny animals create when struggling to escape from the water. Striders may wait for food to drift by or may actively "skate" across the water searching for food. Communication between individual striders is done by sending "tapping" vibrations along the water surface. Water striders breed during the spring and early summer. During these seasons it is common to see striders riding "piggy-back" as they mate.

What You Need

For each group of four students:
❑ a light-colored basin for observing the insects
❑ a clear plastic cup
❑ a long-handled net
❑ (*optional*) a hand lens

Observing Water Striders

1. Divide the class into groups of four students each, and establish the limits of the activity site.

2. Announce that students will be exploring the movement and feeding behavior of water striders, one of the predators of mosquitoes and their larvae.

3. Emphasize that these are delicate insects, and their legs can easily be broken off. While stressing the need for gentle handling, show students how to:

a. Slowly approach wary striders.

b. Use an underhand scooping motion to net the striders.

c. Transfer a netted strider into a basin which contains a small amount of water. (Use care, as striders can jump out of either small or large containers if the water level is less than 2" from the top.)

d. Use a hand lens to view the strider.

e. To view the striders from the side, place them in a clear plastic cup which contains a small amount of water.

4. Direct students to do a quick count of the striders in their sample area before attempting to catch one or two for close observation.

5. As students catch striders, circulate among the groups, and encourage close observation by asking questions such as:

- What do their mouths look like?

- How many legs do striders have?

- What parts of their legs do striders place in the water for support?

- Which legs do striders use to move?

- Are striders wet or dry?

6. Students can observe the striders catch and eat their prey by quietly introducing a fluttering moth or mosquito onto the surface of the water near the strider.

GUIDE TO FRESHWATER LIFE

Name: _____

Aquatic Animals

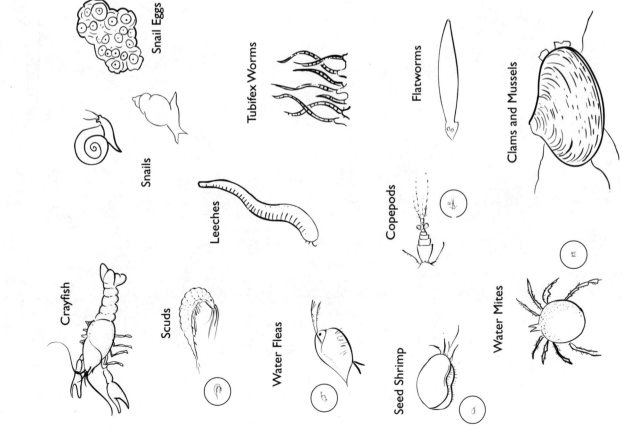

Snail Eggs

Snails

Tubifex Worms

Flatworms

Clams and Mussels

Crayfish

Scuds

Leeches

Water Fleas

Copepods

Seed Shrimp

Water Mites

Fish

Fish spend their lives entirely in water, breathing by means of gills. They have fins, and their streamlined bodies are usually covered with scales.

Amphibians

Amphibians begin life with gills in water and later develop lungs. Their skin is thin, scaleless, smooth or warty, and usually moist. Frogs, toads, and salamanders belong in this group.

Frogs

Frogs are smooth skinned with long, powerful hind legs. Tree frogs have toes with enlarged tips.

Toads

Toads possess a warty skin, large neck glands, and are rarely found moving about during the day. Toads have shorter back legs than frogs have.

Tadpoles

Tadpoles are the well-known larvae of frogs and toads. They are completely aquatic.

Salamanders

Salamanders also include newts. They have lizardlike bodies but lack claws.

Salamander Larvae

These larvae are completely aquatic and possess external gills, which can help you distinguish these larvae from tadpoles.

Toads

Frogs

Tadpoles

Salamanders

Salamander Larvae

Aquatic Insects

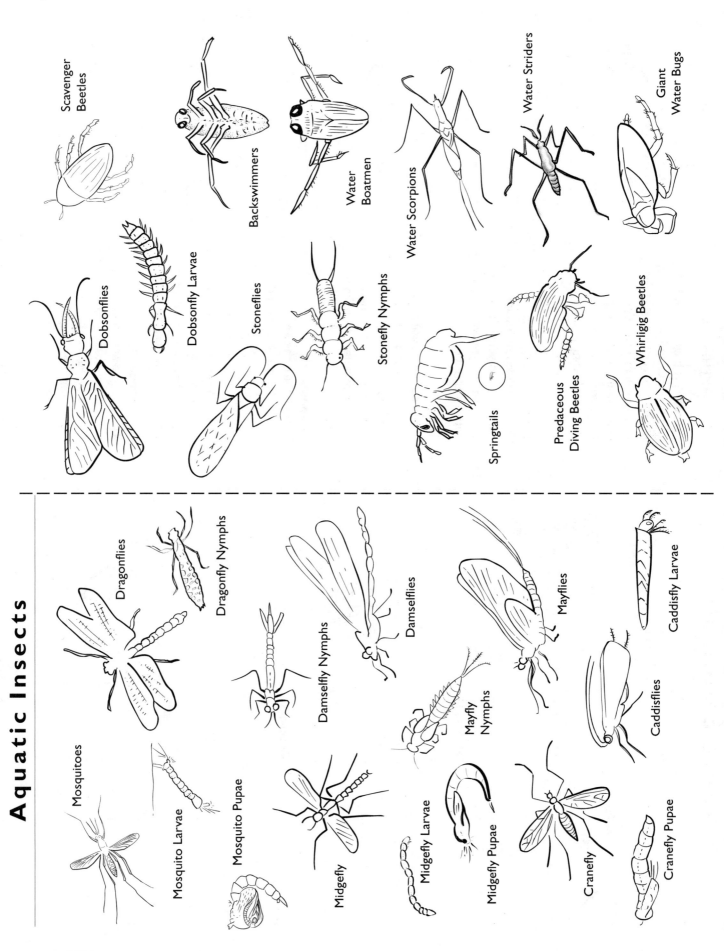

Scavenger Beetles

Backswimmers

Water Boatmen

Water Scorpions

Water Striders

Giant Water Bugs

Dobsonflies

Dobsonfly Larvae

Stoneflies

Stonefly Nymphs

Springtails

Predaceous Diving Beetles

Whirligig Beetles

Dragonflies

Dragonfly Nymphs

Damselfly Nymphs

Damselflies

Mayflies

Caddisfly Larvae

Mayfly Nymphs

Caddisflies

Mosquitoes

Mosquito Larvae

Mosquito Pupae

Midgefly

Midgefly Larvae

Midgefly Pupae

Cranefly

Cranefly Pupae

Behind the Scenes

*The following information is provided as background. This section is **not** meant to be read out loud to or distributed directly to your students. It is intended to provide necessary and concise background for you in presenting the activities and responding to student questions. Please see the "Resources" section for books and other materials that will help you and your students delve more deeply into ponds.*

Maintaining Healthy Aquatic Habitats

The simplest way to maintain water quality is to replace about 20% of the water every three or four weeks with clean, dechlorinated water. The replacement water should be the same temperature as the water in the container.

Ammonia and nitrites from the fishes' waste are the most common pollutants that can cause problems early in the life of the aquarium, before the bacteria that normally absorb these chemicals have had a chance to become established. These pollutants tend to peak several days after the fish are first added. If the fish have clamped fins held tight to the body and become sluggish, replace 20% of the water in the aquarium.

Keep the lids on the habitats when not in use to minimize any dust, or stray drops of soap, paint or other pollutants that may accidentally get into the water. If you think your custodian may use insecticide in your classroom at any time during the unit, be sure to arrange to cover the habitats completely or remove them.

You may want to purchase the easy to use aquarium store test kits for pH, ammonia, and nitrites, so that students can monitor changes in the quality of the water.

What If Animals Die?

On a reassuring note, the plants and animals used in these activities are fairly hardy and have been selected for their tolerance of a wide range of conditions. But be prepared for some attrition. **Remove dead animals from the water as soon as possible to prevent bacterial decay from fouling the water and killing other organisms.**

The most common time to lose large numbers of animals is soon after they're delivered to your classroom. Aquatic animals often weaken or die during the delivery process or while trying to adapt to new conditions.

The best way to avoid the extra expense and disappointment of a big die-off is by careful planning and scheduling. Follow the directions in the "Getting Ready" sections of the guide to prepare for the arrival of animals. Especially for the Tubifex worms, fish, and mosquito larvae, try to minimize the time they are in the transfer containers. Ask your local aquarium store what day of the week fresh animals arrive there, and try to pick them up and transfer them promptly to your holding tank. If you're using a mail order company, it's worth checking and double checking to be sure they will arrive when you need and expect them. Get a parent or other adult volunteer to help with this if possible.

Once transferred and established in the students' habitats, the animals' survival rate is usually pretty high. However, students may be devastated by the occasional loss of a fish or snail in their care. Help them understand that death is a natural process that is common in aquatic habitats. Point out that the life spans of these small animals are brief, and use the incident as a way to learn more about the importance of the animal in the food web.

What Should I Do with the Animals When the Unit is Finished?

DO NOT RELEASE CLASSROOM ORGANISMS INTO WILD ECOSYSTEMS, and don't flush them down the toilet. The introduction of alien species into local habitats has been responsible for the decline of many native organisms. Alien species will consume native species and compete for the same food and shelter.

After your students have completed the aquarium habitat unit, you have several options for the disposition of the living organisms.

1. Set up a large classroom aquarium. All of the organisms from the eight student tanks can be combined together into a single ten gallon tank. Installing a filter will help keep the tank clean. Some filters contain special biological substrates that contain organisms that help break down waste products. The mosquito fish will breed and produce young. As the tank becomes overcrowded with new fish, the young may be consumed by the largest adult fish.

2. Return organisms to local agencies or stores. Your local mosquito abatement district will usually accept returned mosquito fish. Local aquarium stores may accept Elodea, aquatic snails, and fish—and may even provide you with store credit.

3. Introduce organisms into backyard ponds. Parents and local neighbors with outdoor ponds may be very happy to introduce your aquarium organisms into their artificial, backyard pond. The mosquito fish will provide the added benefit of consuming mosquito larvae. Backyard ponds can be a large source of urban mosquitoes.

Basic Information on Organisms

Algae

Algae are green plants that will eventually cover the rocks, sand, and sides of the tanks and cause the water to be greenish. Algae are beneficial as they can produce 60% of the oxygen in the tank during daylight hours. At night when photosynthesis is not occurring, the algae consume oxygen like the aquatic animals.

There are different types of algae. Filamentous (long and stringy) algae can indicate an excess of organic debris. A small amount, however, is a good source of food for the fish and can provide shelter for any young which may be born during the unit.

Unicellular algae, which is microscopic, turns the water green. As with filamentous algae, an excessive amount can indicate a high level of organic decomposition and a low level of oxygen.

As food for the snails, mosquito larvae, and even for the mosquito fish, algae is an important part of the model habitats.

Elodea Water Plants

Elodea grows entirely submerged as a loosely rooted or free-floating plant. The branched stems are crowded with green, translucent, narrow leaves arranged in whorls of three or more. Elodea spreads with amazing speed and may literally fill up a pond or slow stream and crowd out other plants.

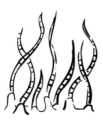

Tubifex Worms

Tubifex worms (scientific name *Tubifex tubifex*) are slender, segmented, reddish worms about an inch and a half long, that live in the soft bottom mud of ponds and slow moving rivers and streams. Because they are easy to keep and breed, Tubifex have become a mainstay of the aquarium supply industry, providing an inexpensive live food for many fish. They tolerate low levels of oxygen in the water, which make them well adapted for recycling the waste at the bottom of aquariums.

Students will notice that the worms immediately cluster into a ball when they're in a container that lacks a bottom substrate. This behavior may facilitate breeding, and may be an adaptation to avoiding predators. Students have commented that the mass of worms is harder for the fish to attack than the single vulnerable Tubifex wiggling its way towards the protective cover of the pebbles on the bottom.

Water Snails

Snails usually have a single-coiled shell that may be rounded, flattened (like a wheel) or long and pointed. Some shells coil to the right and others are "left handed." Snail eggs look like small, transparent sheets or globules of jelly. If you look closely, you can see the developing snails (dark spots). Snail eggs are often attached to water plants, the sides of the aquarium, sticks, stones, or floating objects.

Mosquito Fish

The mosquito fish, *Gambusia affinis,* is native to the Southern United States from Florida to Texas. This small hardy fish has been introduced to ponds and wetlands throughout the world because of its great appetite for mosquito larvae. An adult Gambusia can consume its own weight in mosquito larvae every day! It is ironic that this fish which is so useful in controlling mosquitoes has a name derived from the Spanish slang word *gambusino,* which means "worthless."

Unfortunately Gambusia often cause problems in their non-native habitats by decimating native fish and insects. Agents of mosquito abatement and vector control agencies are very careful not to introduce Gambusia to habitats known or thought to contain endangered or threatened species.

Gambusia will tolerate a wide variety of water conditions, with temperatures ranging from just above freezing to about 85° F. Males grow to 1 ¼ inches (3 cm), and the females to 2 ¼ inches (6.5 cm). Females can produce up to four broods per year and bear live young that grow very rapidly. The fish prefer live food like mosquito larvae and small pond organisms. The fish are very aggressive towards tank mates, frequently attacking their fins. The shelter used in the aquarium provides a place for smaller fish to hide.

Life Cycle of the Mosquito

Mosquitoes belong to the order of insects known as "flies," and have one pair of transparent wings, fringed with tiny colorful scales and hairs. The adult females lay their eggs either on the surface of water, or in a place that will later be covered by water. There are more than 2000 different kinds, or species, of mosquitoes spread across the wetlands of the world. About 20% of these have played a role in spreading diseases like malaria, yellow fever, dengue, and encephalitis, even affecting the outcome of wars and colonial expansion.

Most mosquitoes are equipped with a long piercing and sucking tube for obtaining their food. Males feed on plant juices. It is only the females of certain species that must suck blood for their eggs to develop. They bite because they lay their eggs in habitats with little nutritious food and need the protein from blood to help the eggs and larvae grow. Each kind of female mosquito prefers a certain kind of food. Certain ones will feed only on plant juices or on cold-blooded animals such as frogs. Other mosquitoes prefer birds. Still others prefer to suck the blood of large animals such as horses and humans.

Female mosquitoes hover in swarms near the place where they lived as larvae. The males are attracted by the humming wings of the females and fly into the swarm, where they mate. Once the eggs are fertilized by the sperm of the males, the female lands on the surface of still water, or on vegetation and lays clusters of eggs called "rafts." From the time they are laid, a host of microorganisms and nematode worms feed on the eggs. After a few days, tiny larvae emerge from the eggs and live for the next several days in shallow water. These larvae are commonly called "wigglers" because of the wiggling motion they make as they swim. These wigglers hang downward from the surface film and breathe by means of gills and an air tube near the rear end of the body. The wigglers are very active and feed on microorganisms and algae. The wigglers are eaten by many insects, including dragonfly and damselfly

Mosquito Life Cycle

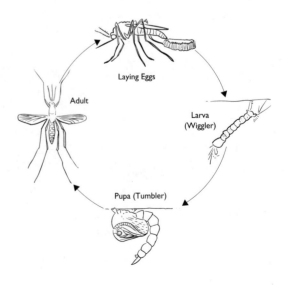

Laying Eggs

Adult

Larva
(Wiggler)

Pupa (Tumbler)

nymphs, back swimmers, giant water bugs, predaceous diving beetles, fish, and many other aquatic predators.

After several days, the larvae metamorphose into non-feeding pupae that look like large commas, resting at the surface of the water. They breathe through two tubes near the head, and, if disturbed, tumble away using their flap-like tails, giving rise to the nickname "tumbler." Within a few days, the back skin of the pupae splits allowing the winged adult mosquito to pull itself onto the surface of the water where it floats momentarily before taking wing.

Starting a Mosquito Larvae Culture

Mosquito larvae can be purchased from some scientific supply companies, but an educational and inexpensive alternative is to start your own culture with the aid of house mosquitoes and fish-pond mosquitoes living in your area. The weather must be generally mild and nighttime temperatures must be above freezing. About a month before starting the unit, set a bucket, half full of dechlorinated water, in a sunny outdoor location to attract female mosquitoes to lay eggs. If possible, add a cupful of pond water containing algae to serve as food for the larvae.

Avoid placing the bucket where it will catch a lot of debris from surrounding vegetation. Do not put the bucket anywhere near resinous plants such as Laurel, Eucalyptus, Oleander, Redwood, or other "anti-mosquito" plants. These plants produce resins that are toxic to many aquatic organisms.

Within about a week, female mosquitoes will land on the surface of the water and lay their small rafts of eggs. The life cycle of the mosquito depends on the species of mosquito, the temperature of the water, and the food supply of microorganisms living in the water. Pollen and algae from the surrounding environment will fall into the water and provide food for the mosquito larvae. On average, the eggs will hatch within a day or so and the larvae will take a little more than a week to develop into pupae. During the culturing period, maintain the water level and keep the bucket free of predators such as back swimmers and water striders which can fly in from nearby wetlands.

Use a dip net to transfer the larvae into a smaller class-room tank with a lid. When the pupae metamorphose at the surface, the adult mosquitoes will pull free of the water and fly to the sides of the tank. To prevent their escape, don't forget to keep the tank covered. When you have gathered enough larvae for your classroom activities, empty the water from the culture bucket so you don't accidentally contribute to the mosquito population in your area.

Mosquito Control

Mosquitoes are pests throughout most tropical and temperate regions of the world. The small amount of anticoagulation enzyme that they inject into their victim in the process of obtaining blood, often produces a very bothersome itching and swelling. Add to this the fact that mosquitoes are the carriers of a number of the world's most debilitating diseases, (malaria, yellow fever, dengue, and sleeping sickness, among others), it is no wonder that they are some of the least appreciated life forms on earth. Although most of the disease carrying mosquitoes are found in the warmer regions, great clouds of mosquitoes are found across the marshes and bogs near the Arctic Circle and along the coastal wetlands of the temperate regions of the world.

The animal world abounds with adaptations for surviving biting insects, including thick hide, swishing tails, and mud baths. Every human culture with mosquito neighbors has developed techniques for thwarting these bothersome insects, and their struggles are depicted in drama, stories, and art. People avoided building towns near wetlands, and have developed many lotions and body coatings, such as clay and animal fat, to protect themselves. Some cultures encourage the natural predators of mosquitoes. In the tropics it is common for people to have gecko lizards and frogs in their homes. Some European communities have built elaborate bird nest boxes for swifts and martins. Dragonflies and bats are often revered for their mosquito eating habits.

In areas of extensive wetlands, populations of mosquitoes reach staggering proportions. Efforts to control this pest have often resulted in drained and destroyed bogs and swamps, or oil and poison spread widely over breeding areas. These efforts can also destroy the mosquitoes' natural enemies, thus compounding the problem. Modern technologies and use of poisons have often resulted in resistant strains of mosquitoes and resistant forms of the diseases they inadvertently carry.

In recent years, scientists have been experimenting with biological mosquito control techniques that make use of natural diseases and predators (such as Gambusia) to reduce mosquito populations. Many pest control agencies now use biological control methods as an alternative to chemicals. With this method of control, natural enemies regulate the population size of pest organisms.

Biological control is the process of using natural controlling organisms to reduce the numbers of a pest organism. The agent of control can be a disease, a parasite, or a predator. For example, fish and aquatic insects consume huge quantities of mosquito eggs and wigglers. Thousands of species of birds and bats specialize in capturing mosquitoes on the wing. The eggs of microscopic nematode worms are being sprayed over flooded areas. When the nematodes hatch, they eat the eggs and very young larvae of mosquitoes. Identifying controlling organisms and enhancing their populations in heavily infested mosquito areas are the practices of biological pest control.

The majority of biological control efforts focus on the mosquito larvae, because if the larvae are killed the problem of the adults biting and spreading disease is eliminated.

The following are a few examples of modern mosquito abatement efforts:

- Naturally occurring bacteria that produce spores that are toxic to the larvae of mosquito and black fly larvae are sprayed on flooded lands.

- An insect growth regulator that is specific to mosquitoes is being applied to flooded areas to prevent the larvae from developing.

- Mosquito fish are no longer released into natural streams, lakes and wetlands, however, they are stocked in flood control canals and artificial water sources. In Berkeley, California, there are more than 1200 backyard ornamental ponds. Local mosquito control agencies provide the hardy mosquito fish free of charge to homeowners who would like some for their pond.

- Abatement programs eliminate standing water from ditches and roadsides by creating drainage pathways for the trapped water.

Mosquito-Borne Diseases

One of the great advances in the history of medicine was made by a government commission sent to investigate yellow fever in Cuba in 1900. Under the direction of Dr. Walter Reed, the study proved that the mosquito *Aedes aegypti* was the carrier of yellow fever. The commission found that if the mosquitoes were kept away from patients with the disease, there was no danger of the disease spreading to others. The mosquito had to bite a person with yellow fever before it could spread the germ by biting another person. This knowledge revolutionized the treatment and prevention of insect borne diseases including malaria and dengue. Early recognition of the diseases and immediate isolation of the infected patients became effective tools in preventing epidemics.

Aedes aegypti is adapted for feeding exclusively on human blood and seldom is found more than a half a mile from where people are living. The female mosquitoes prefer to lay their eggs in barrels and cisterns, just above the water line. An understanding of the life cycle of this mosquito has enabled people to eradicate it from their neighborhoods. In areas where rainfall is frequent, people are careful to clean up discarded containers such as barrels and cans. Water troughs for livestock are flushed regularly and people are always on the lookout for potential breeding areas.

Some Sources for Key Items

The "Key Items and Alternatives" chart on pages 8–9 provides an overview of important items you'll need to arrange for before beginning the unit, plus a list of alternatives for items that may be expensive or hard to find in your area.

Listed below are some sources for materials or organisms you may want to order. All companies listed here ship anywhere in the United States. Prices and other information are of course subject to change, and *shipping costs are not included here.* The addresses for the companies listed below are at the end of this section.

▶ Aquarium tanks for student groups

You'll need one plastic tank and lid for each student group. Tanks should be flexible with clear, undistorted sides. (Rigid ones lead to cracking and breakage.) Ask for the lids with small, round holes, rather than elongated holes. Those listed below are easy to clean and nestable for storage, and can be used year after year, either as aquaria or terraria.

Acorn Naturalists

Cat. No. T-6271	1 ½ gallon Flex-Tank	$4.60
Cat. No. T-6272	Flex-Cover (lid with holes for ventilation)	$1.80

Flinn Scientific

Cat. No. FB0269	1 ½ gallon flexible plastic aquarium with screen cover	$9.55
Cat. No. FB0270	1 ½ gallon flexible plastic aquarium with dome cover (with dial-type ventilator)	$12.25

Frey Scientific

Cat. No. S15567	1 ½ gallon plastic stacking aquarium	$6.45
Cat. No. S15568	fitted cover with ventilation air control dial	$4.85

Nasco Science

Cat. No. SB19273M	1 ½ gallon Flex-Tank with Flex-Cover (lid with holes for ventilation)	$6.25 each including lid $5.50 each for 10 or more $4.90 each for 50 or more

Cat. No. WL8329-10   1 ½ gallon flexible $9.90
                      plastic aquarium with
                      cover (with adjustable
                      ventilator)

▶ Dechlorinating/dechloraminating liquid

Whether your water is treated with chlorine or chloramine, you can remove it instantly with this liquid. Read about dechlorination on page 15 and use whatever your aquarium store recommends, or order:

Aquatic Eco-Systems, Inc.

Cat. No. CL16   16 oz. tap water $4.99
                     conditioner

▶ Sand and gravel

Light-colored sand and gravel can be purchased from aquarium stores as well as from construction suppliers, sand and gravel companies, garden centers, hardware stores, or home centers. If obtained from aquarium stores, sand and gravel are usually clean, but if purchased from other sources they may need to be rinsed to remove excess dust or debris. You can also order sand and gravel from the following sources (remember, though, that shipping may cost as much as the item itself):

Nasco Science

Cat. No. S09940M   10 lbs. white silica sand $2.85

Cat. No. S09944M   10 lbs. natural-colored $3.20
                     flint gravel

VWR Scientific Products/Science Education (also known as Sargent Welch)

Cat. No. WLS-1050-20A   5 lbs. white $4.60
                        silica sand

Cat. No. WLS-1053-B   5 lbs. natural color $4.40
                        gravel

▶ Tubifex worms

Most aquarium stores carry live Tubifex worms. Look in the yellow pages under "Aquariums & Aquarium Supplies" or "Tropical Fish" to find stores in your area. You can also order from the following sources:

Carolina Science and Math

Cat. No. P7-L396   Tubifex worms $5.70 for enough for 30
                                                    students

Novalek, Inc.

Cat. No. 51125	Tubifex worms	$ .90 for a 30 ml (1 oz.) bag
Cat. No. 51225	Tubifex worms	$1.85 for a 150 ml (5 oz.) bag

VWR Scientific Products/Science Education (also known as Sargent Welch)

Cat. No. WL50617	Tubifex worms	$6.99 for enough for 25–50 students

▶ **Daphnia**

Daphnia are a freshwater crustacean that can serve as an alternative to Tubifex worms as a food source for the fish. If you obtain some natural pond water, most likely it will have tiny Daphnia flitting about in it. You may be able to purchase Daphnia at your local aquarium store, or order them from the following sources:

Carolina Science and Math

Cat. No. P7-L565	Daphnia	$6.00 for enough for 30 students

Flinn Scientific

Cat. No. LM1107	small feeder Daphnia	$5.75 for enough for 30 students
Cat. No. LM1108	small feeder Daphnia	$17.25 for enough for 100 students

Frey Scientific

Cat. No. S05685	Daphnia	$5.70 for 25
Cat. No. S11266	Daphnia	$12.95 for 100

Nasco Science

Cat. No. LM00754M	small feeder Daphnia	$10.75 for 100

Novalek, Inc.

Cat. No. 51155	Daphnia	$ .90 for a 30 ml (1 oz.) bag
Cat. No. 51255	Daphnia	$1.85 for a 150 ml (5 oz.) bag

VWR Scientific Products/Science Education (also known as Sargent Welch)

Cat. No. WL50716	Daphnia	$6.89 for enough for 25–50 students

▶ Pond snails

You can obtain freshwater (aquatic) snails from your local aquarium store, or order from the following sources:

Carolina Science and Math

Cat. No. P7-L489A	small pond snails	$5.89 for 12
		$11.00 for 25
Cat. No. P7-L489	medium pond snails	$7.50 for 12
		$12.75 for 25

Flinn Scientific

Cat. No. LM1106	pond snails	$7.70 for 12

Frey Scientific

Cat. No. S11253	pond snails	$6.65 for 12

Nasco Science

Cat. No. LM00035M	large pond snails	$6.00 for 10

Allow one and a half weeks to process the order.

VWR Scientific Products/Science Education (also known as Sargent Welch)

Cat. No. WL50675	small pond snails	$6.99 for 12
Cat. No. WL50677	large pond snails	$6.99 for 12

▶ Gambusia (mosquito fish)

In many parts of the United States, Gambusia are used as a biological control for mosquitoes. Many local and regional mosquito abatement or vector control agencies have free Gambusia available for local residents. On the next page is a *partial* list of agencies that, as of this publication, provide free Gambusia for mosquito control to residents of their regions.

In your search for free Gambusia, a phone book can be a valuable tool. Using the Government pages, you may need to begin with your local public or environmental health department. They may refer you to a mosquito abatement or vector control agency whom you could then call to determine the availability of free Gambusia. You may also try your state fish and game agency or a local pond supply or nursery as suppliers of Gambusia.

Also, refer to the Mosquito-related Internet Sites listed in the "Resources" section for several regional, state, and local mosquito and vector control agencies.

California

Alameda County Mosquito Abatement District
23187 Connecticut Ave.
Hayward, CA 94545
(510) 783-7744

Contra Costa Mosquito and Vector Control
District
155 Mason Circle
Concord, CA 94520
(925) 685-9301

Goleta Valley Vector Control District
P.O. Box 1389
Summerland, CA 93067
(805) 969-5050

Los Angeles County West Vector Control District
6750 Centinela Ave.
Culver City, CA 90230
(310) 915-7370

Marin/Sonoma Mosquito and Vector Control
District
556 N. McDowell Boulevard
Petaluma, CA 94954
(707) 762-2236

Shasta Mosquito and Vector Control District
P. O. Box 990331
Redding, CA 96099-0331
(916) 365-3768

New Mexico

Albuquerque Environmental Health Department
Insect and Rodent Control Division
P.O. Box 1293
Albuquerque, NM 87103
(505) 873-6613

Chaves County Vector Control Program
P.O. Box 1817
Roswell, NM 88202
(505) 624-6610

Colfax County Vector Control Program
P.O. Box 1498
Raton, NM 88740
(505) 445-2906

Eddy County Vector Control Program
410 E. Derrick Rd.
P.O. Box 1139
Carlsbad, NM 88221
(505) 885-4835

Oregon

West Umatilla Vector Control
Rt. 1, Box 1961-A
Hermiston, OR 97838
(541) 567-5201

Utah

Utah Mosquito Abatement
2020 N. Redwood Rd.
Salt Lake City, UT 84116
(801) 355-9221

Davis County Mosquito Abatement
85N 600W
Kaysville, UT 84037
(801) 544-2864

Washington

Benton County Mosquito Control District
6174 W. Van Giesen
West Richland, WA 99353
(509) 967-2414

If you are unable to find free Gambusia, you may be able to buy them at a pond supply store or nursery, or you could order them from these sources.

Carolina Science and Math

| Cat. No. P7-L1379 | Gambusia | $8.25 for 12 |
| | | $27.05 for 50 |

Natural Pest Controls

| | Gambusia | $20.00 for 25 |

▶ Mosquito larvae and eggs

You can cultivate your own mosquito larvae (see page 84) or get them from a nearby pond. Alternatively, you may want to order them from the following source:

Carolina Science and Math

Cat. No. P7-L999	mosquito egg raft	$9.75 for about 100
Cat. No. P7-L1001	lab-reared mosquito larvae	$11.00 for 25
		$38.95 for 100

Addresses of suppliers listed above

Acorn Naturalists
17300 East 17th St., #J-236
Tustin, CA 92780
(800) 422-8886
fax (800) 452-2802
e-mail: Acorn@aol.com
www.acornnaturalists.com

Aquatic Eco-Systems
1767 Benbow Court
Apopka, FL 32703-7730
(407) 886-3939
(800) 422-3939
fax (407) 886-6787
e-mail: aes@aquaticeco.com
www.aquaticeco.com

Carolina Science and Math
2700 York Rd.
Burlington, NC 27215-3398
(800) 334-5551
fax (800) 222-7112
e-mail: carolina@carolina.com
www.carolina.com

Flinn Scientific
P.O. Box 219
Batavia, IL 60510-0219
(800) 452-1261
fax (630) 879-6962
e-mail: flinnsci@aol.com

Frey Scientific
100 Paragon Parkway
Mansfield, OH 44903
(888) 222-1332
fax (888) 454-1417
www.beckleycardy.com

Nasco Science
901 Janesville Ave.
Fort Atkinson, WI 53538-0901
or
4825 Stoddard Rd.
Modesto, CA 95356-9318

(800) 558-9595
Wisconsin fax (920) 563-8296
California fax (209) 545-1669
e-mail: info@nascofa.com
www.nascofa.com

Natural Pest Controls
8864 Little Creek Dr.
Orangevale, CA 95662
(916) 726-0855

Novalek, Inc.
2242 Davis Ct.
Hayward, CA 94545-1114
(510) 782-4058
fax (510) 784-0945

**VWR Scientific Products/
Science Education
(also known as Sargent Welch)**
P.O. Box 5229
Buffalo Grove, IL 60089-5229
(800) 727-4368
fax (800) 676-2540
e-mail:
sarwel@sargentwelch.com
www.sargentwelch.com

Resources

Related Curriculum Material

There are a wealth of curriculum materials relating to ecology and the environment. To explore the many units available, we recommend that you check out resources such as the National Science Resources Center publication entitled *Resources for Teaching Elementary School Science,* National Academy Press, Washington, 1996, and *NSTA Pathways to the Science Standards,* Lawrence F. Lowery, editor, National Science Teachers Association, Arlington, Virginia, 1997. Both of these books include listings and concise annotations of life science, ecology, and environmental units that would nicely complement or extend the activities in *Aquatic Habitats.* You will also find many curriculum materials recommended in publications of leading environmental education organizations and others highlighted on the Internet, as well as many related units on insects and plants. We welcome your suggestions.

Aquatic Education Activity Guide
by Project WILD-Aquatic
Project WILD National Office
5430 Grosvenor Lane #230
Bethesda, MD 20814

This activity guide is not for sale; rather it is available to teachers and other educators who have attended professional development sessions offered by coordinators in each state. For information on your state, call the Project WILD National Office at (301) 493-5447 or visit their web site at eelink.umich.edu/wild/

Pond and Brook: A Guide to Nature in Freshwater Environments
by Michael J. Caduto
University Press of New England, Hanover, New Hampshire. 1990

Contains information and hands-on activities for teachers and students to investigate freshwater environments. Designed for the amateur naturalist, this is a valuable tool for teaching about the unique properties of water, the basic principles vital to understanding aquatic life, and the origin of freshwater habitats.

Project WET Curriculum and Activity Guide

by Project WET Staff
201 Culbertson Hall
Montana State University
Bozeman, Montana 59717-0570

Developed for K–12 students, this guide is a collection of innovative, water-related activities that are hands-on and easy to use, and incorporate a variety of learning formats. The relationship of people to water is a major theme of the guide which also addresses water's chemical and physical properties, quantity and quality issues, aquatic wildlife, ecosystems, and management strategies. The activities in the guide promote critical thinking and problem-solving skills and help provide students with the knowledge and experience they will need to make prudent decisions regarding water resource use. For more information contact Project WET at the address above, or call (406) 994-5392.

WOW! The Wonders of Wetlands: An Educator's Guide

The Watercourse
P.O. Box 170575
Montana State University
Bozeman, MT 59717

This is a comprehensive guide for developing a wetlands study program appropriate for grades K–12. Each activity features clear headings detailing grade levels, skills, materials, field or lab procedures, and appropriate assessment techniques. The guide also gives details on how to plan and develop your own wetland habitat.

The following **Outdoor Biology Instructional Strategies (OBIS)** aquatic activities, developed at the Lawrence Hall of Science, are available from Delta Education (1-800-258-1302):

Attract a Fish	Water Breathers
Can Fishing	Water Holes to Mini-ponds
Crawdad Grab	Water Snails
Damsels and Dragons	Water Striders
Habitats of a Pond	What Lives Here?
Too Many Mosquitoes	

The **Full Option Science System (FOSS),** developed at the Lawrence Hall of Science, includes an excellent module that relates to the activities in this GEMS guide. The **Environments** module, for Grades 5 and 6 (6 activities) provides structured investigations in both terrestrial and aquatic systems to develop concepts of environmental factor, tolerance, environmental preference, and environmental range. FOSS modules are available from Delta Education, 5 Hudson Park Drive, Hudson, New Hampshire 03051-0915, (800) 258-1302.

Mosquitoes in the Classroom: A Teacher Resource Guide and Classroom Curriculum
by Frances J. Spray and the
University of Wisconsin-Madison Medical Staff
Kendall/Hunt Publishing, Dubuque, Iowa, 1995.

In this guide, mosquitoes are used to teach a variety of subjects— from basic science and health to music, art, math, geography, and more. It's a complete curriculum unit designed for all grade levels, with a special focus on elementary and middle grades.

The **Aquatic Outreach Institute** conducts *Kids in Creeks, Kids in Marshes, Kids in Gardens,* and *Watching Our Watersheds* workshops for educators in Alameda, Contra Costa, and Marin counties (in the San Francisco Bay Area). These workshops offer resources and activities to assist educators in teaching about the importance of protecting local watersheds. Alumni of these workshops can apply for grants to provide supplies and equipment for creek or marsh studies for their students. The Aquatic Outreach Institute also sponsors an annual conference focused on watershed education. For more information contact:

> Dede Sabbag, Education Program Coordinator
> Aquatic Outreach Institute
> 155 Richmond Field Station
> 1327 South 46th Street
> Richmond, CA 94804
> (510) 231-5784

Catalogs of Relevant Material

Acorn Naturalists
17300 East 17th Street, #J-236
Tustin, CA 92780
(800) 422-8886
www.acornnaturalists.com

This catalog is filled with resources for exploring the natural world. It contains a myriad of materials such as field guides, equipment, puppets (including a mosquito finger puppet), and a large variety of books—including several on understanding and developing environmental education programs.

Environmental Media Corporation
P.O. Box 99
Beaufort, SC 29901-0099
(800) 368-3382
www.envmedia.com

This catalog contains many environmental education resources
for the classroom and the community. It includes books, videos,
CD-ROMs, and equipment. The catalog is conveniently arranged
by topics such as biodiversity, forests, microlife, and environmental stewardship.

Field Guides

A Field Guide to the Insects of America North of Mexico, Donald
Borror and Richard White, Houghton Mifflin, Boston, 1970.

Insects: A Guide to Familiar American Insects, Herbert S. Zim and
Clarence Cottam, Golden Press, New York, 1961.

*The National Audubon Society Field Guide to North American Insects
& Spiders,* Lorus and Margery Milne, Alfred A. Knopf, New York,
1980.

*Pond Life: A Guide to Common Plants and Animals of North American
Ponds and Lakes ,* George K. Reid, Golden Press, New York, 1967.

Books

General Pond Life

The Hidden Life of the Pond, David M. Schwartz, Crown Publishers,
New York, 1988.

Photographs and text introduce the animals, insects, and plants
in a pond.

One Small Square: Pond, Donald M. Silver, W. H. Freeman, New
York, 1994.

Contains descriptions of the variety of life at different levels of a
pond—in the air above it, at the surface of the water, in the water,
and deep in the pond—as well as at different times of the day or
year. Contains activities to explore a small square of a pond.

The Pond: The Life of the Aquatic Plants, Insects, Fish, Amphibians, Reptiles, Mammals, and Birds that Inhabit the Pond and its Surrounding Hillside and Swamp, Jack Samson, Knopf, New York, 1979.

Describes the interdependency and seasonal changes in the plant and animal life in and around a pond.

Pond & River, Steve Parker, Knopf, New York, 1988.

An Eyewitness book about the range of plants and animals found in fresh water throughout the year. Examines the living conditions and survival mechanisms of creatures dwelling at the edge of the water, on its surface, or under the mud.

Pond Life, Barbara Taylor, Dorling Kindersley, New York, 1992.

Examines the variety of life found in ponds, including the common newt, stickleback, and great diving beetle.

PondWatchers Guide to Ponds and Vernal Pools of Eastern North America, the Massachusetts Audubon Society, Educational Resources, 208 South Great Road, Lincoln, MA 01773; (617) 259-9500.

A two-sided, fold-out, laminated guide with color illustrations and descriptions of many pond inhabitants. Also includes information on ponds and vernal pools, energy flow, and seasonal changes.

Puddles and Ponds, Rose Wyler, Julian Messner, Englewood Cliffs, New Jersey, 1990.

Describes in clear text and illustrations the plants and animals that live in and around ponds and other small bodies of water. Includes instructions for simple experiments.

Mosquitoes

Let's Find Out About Mosquitoes, David Webster, Franklin Watts, New York, 1974.

Introduces the physical characteristics and habits of the mosquito.

Mosquito, Oxford Scientific Films, Putnam, New York, 1982.

Discusses the biology and life history of mosquitoes and the dangers of mosquito-borne diseases.

A Mosquito is Born, William White, Jr. and Sara Jane White, Sterling Publishing, New York, 1978.

Describes and illustrates in detail the habitat, physical characteristics, and life cycle of mosquitoes and how they transmit diseases.

Mosquitoes, Dorothy Hinshaw Patent, Holiday House, New York, 1986.

Discusses the mosquito's habits, development, and diseases it carries, as well as ways to control these creatures.

Water Insects

Water Insects, Sylvia A. Johnson, Lerner Publications, Minneapolis, 1989.

Describes the physical characteristics, behavior, and life cycles of some insects that spend most of their lives in the water.

Water Plants

Pond and Marsh Plants, Olive L. Earle, Morrow, New York, 1972.

Describes the characteristics of more than thirty common and rare plants growing in and around ponds and marshes.

Water Plants, Laurence Pringle, Crowell, New York, 1975.

Introduces various plants living in and around ponds, such as cattails, bladderwort, and algae.

General Ecology

Janice VanCleave's Ecology for Every Kid: Easy Activities That Make Learning Science Fun, Janice VanCleave, Wiley & Sons, New York, 1996.

Food Chains/Webs

The Food Chain, Malcolm Penny, Bookwright Press, New York, 1987.

Explains the food chain, links in it, and the food pyramid. Contains examples of food chains in different habitats such as a backyard, a lake, the Arctic tundra, and the African plains.

Pond Life: Watching Animals Find Food, Herbert H. Wong and Matthew F. Vessel, Addison-Wesley Publishing, Reading, Massachusetts, 1970.

A description of the feeding habits of animals living in and around a pond and how these habits affect the ecology of the area.

Predator!, Bruce Brooks, Farrar Straus Giroux, New York, 1991.

Survival in the wild creates a hierarchy of predators and their prey; this interaction among the animals forms the basis of a complex ecological system known as the food chain.

Who Eats What? Food Chains and Food Webs, Patricia Lauber, HarperCollins, New York, 1995.

Explains the concept of a food chain and how plants, animals, and humans are ecologically linked.

Communities

Small Worlds: Communities of Living Things, Howard E. Smith, Jr., Charles Scribner's Sons, New York, 1987.

Describes a number of small self-contained communities, including a sand dune, tidal pool, old barn, and vacant lot, and examines the ways in which the plants and animals interact.

Video

Pond & River, a CAFE production for BBC Worldwide Americas, Dorling Kindersley Vision in association with Oregon Public Broadcasting, 1996.

A mixture of animation and live footage serves as a background for a narrated introduction to the biological life of rivers and ponds and their influence on the history of human beings.

Mosquito-related Internet Sites

On-line mosquito curriculum

- Neato Mosquito: An Elementary Curriculum Guide
  www.cdc.gov/ncidod/dvbid/arbor/neato.pdf

Mosquito Abatement and Vector Control Associations

• American Mosquito Control Association (AMCA)
  www.mosquito.org

• PestWeb (a professional pest control industry site)
  pestweb.com

• Vector Associations and Control Districts
  www.vectorbytes.com/vassociations.html

• Northeast Mosquito Control Association, Inc. (for MA, ME, VT, RI, NH, NY, NJ, and CT)
  www.nmca.org

• Northwest Mosquito and Vector Control Association (for AK, ID, MT, OR, and WA as well as Alberta, British Columbia, and Saskatchewan)
  www.nwmvca.org

• New Jersey Mosquito Control Association
  www-rci.rutgers.edu/~insects/njmca.htm

• New Jersey Mosquito Control Association's "mosquito links" site
  www-rci.rutgers.edu/~insects/links.htm

• Florida Mosquito Control Association
  www.famu.edu/mls/fmca.htm

• Louisiana Mosquito Control Association
  user.maas.net/~cpmc

• Utah Mosquito Abatement Association
  www.umaa.org

• Alameda County (California) Mosquito Abatement District
  mosquito.lanminds.com

• Internet Resources for the Mosquito and Vector Control Associations of California
  mosquito.lanminds.com/directory/Directory.html

• Listing of California Mosquito and Vector Control Agencies
  mosquito.lanminds.com/directory/list.html

• Metropolitan Mosquito Control District (Minneapolis/St. Paul, MN)
  www.mmcd.org

Self-contained Aquarium Ecosystems

Do-Little® Aquarium Kit and
The Kids Do-Little® Mini Aquarium Kit

Manufactured by BioSand International, Inc., these aquariums are miniature aquatic environments that need no pump or filter. The inhabitants of the aquarium (fish, plants, and bacteria) live for the benefit of each other in a continuous cycle. Vital to the system are the BioSand® Beads—a natural product made up mostly of silica dioxide "sand" manufactured into tiny porous spheres—which provide a place for bacteria to live, grow, and multiply. The bacteria thrive on waste given off by the fish and plants in the aquarium. The plants benefit in this cycle by the food conversion or waste produced by the bacteria. Healthy, growing plants produce vital oxygen for the fish through photosynthesis. BioSand® Liquid, a mineral enriched water, provides the aquarium with new "families" of bacteria and vital minerals with every partial water change.

The aquarium are available in a variety of shapes and sizes, including a plastic container for children. Look for them at your local aquarium or pet store, or contact the manufacturer at the address below. Their web page includes a curriculum page for teachers which lists on-line resources for a thematic unit on the ocean plus links to other science curriculum sites on the Internet.

    BioSand International, Inc.
    P.O. Box 696
    Lansdowne, PA 19050
    (800) 724-6726
    fax (610) 284-6955
    www.biosand.com

Assessment Suggestions

Selected Student Outcomes

1. Students improve their ability to make careful observations and to depict their observations accurately through drawing, writing, and in class discussions.

2. Students are able to articulate that an organism's habitat provides it with the necessities for its existence. They are able to describe the habitat needs of aquatic organisms, such as fish, invertebrates, and plants.

3. Students are able to explain how the structures and behaviors of organisms help those organisms survive. Older students are able to relate these understandings to the concept of adaptation.

4. Students increase their understanding of the concept of an organism's life cycle, and are able to draw or describe the life cycle of organisms in the model habitat, such as the mosquito.

5. Students can describe or draw interactions between organisms. They are able to diagram food webs and to describe roles played by different organisms composing an aquatic food web. Students deepen their understanding of the interconnected nature of their model habitat.

Built-In Assessment Activities

Organisms in the Habitat. In Activities 2–5, students observe the various organisms and their interactions. In class discussions and in student journals/observation sheets they record characteristics and behaviors that help organisms survive and grow. They also observe and record changes in the aquatic ecosystem over the course of the entire unit. Their observations—verbal, pictoral, and written—provide the teacher with a great deal of assessment information. Teachers can see how well students use new vocabulary and how they have grasped major concepts. The growth of student observation and recording skills can be tracked as the unit progresses. (Outcomes 1, 2, 3, 4, 5)

Discussing Structures and Behaviors. In Activity 2, students introduce both Tubifex worms and aquatic snails into their model habitats. The teacher circulates to student groups, encouraging and listening to student observations.

Teacher questions focus on the body structures and behaviors of both organisms, locomotion, what they eat, how they avoid predators, etc. While circulating, teachers can assess student observation abilities, note how well students are cooperating, and gain information on misconceptions students may have about organisms and/or their interactions. (Outcomes 1, 2, 3)

Gambusia. In Activity 3, Gambusia are introduced into the habitat. Teachers can assess the further development of observation abilities as students make a close study of the fish before placing them in the habitat. During the class discussion, teachers can gain insight into how well all students are relating the structures and behaviors of the fish to their survival and how well older students grasp the idea of adaptation. (Outcomes 1, 2, 3)

Food Webs. In Activity 4, following the observation and introduction of mosquito larvae and pupae, there is a class discussion on biological control and food webs. Students' food web diagrams can provide excellent information about student understanding of interactions between organisms, the role of decomposers, and, for older students, nutrient cycles. (Outcomes 4, 5)

Experimental Questions. In Activity 5, students document observations through journal entries and drawings over a period of several additional weeks. Teachers can continue to assess observational and descriptive skills. Then students select experimental questions to investigate and document. Teachers can analyze these student experiments and student documentation of them to see how well students apply skills, concepts, and information gained in the unit to their new investigations.
(Outcomes 1, 2, 3, 4, 5)

Closing Essay. Also in Activity 5, a closing essay or summation in the student journal is suggested after the habitats are dismantled. This can provide information on what students have learned and especially what students consider to be the most important messages of the unit. (Outcomes 1, 2, 3, 4, 5)

Additional Assessment Ideas

Diary of an Organism. Students could write a story or diary from the point of view of one of the organisms, whether it be a plant or animal, Elodea or mosquito,

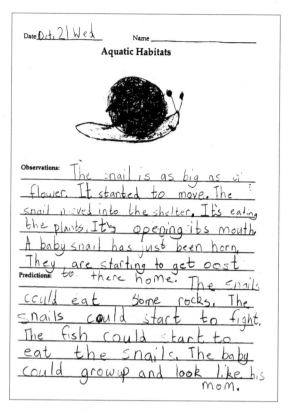

Date Oct. 21 Wed   Name _____

Aquatic Habitats

Observations: The snail is as big as a flower. It started to move. The snail moved into the shelter. It's eating the plants. It's opening its mouth. A baby snail has just been born. They are starting to get oost to there home.
Predictions: The snails could eat some rocks. The snails could start to fight. The fish could start to eat the snails. The baby could grow up and look like his mom.

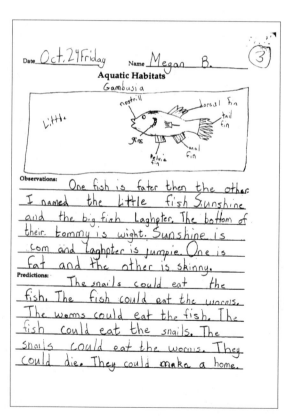

Date Oct. 24 Friday   Name Megan B.   (3)

Aquatic Habitats
Gambusia

Little   nostrill   dorsal fin
                                        tail fin
                    gills
                        pelvic   anal fin
                        fin

Observations: One fish is fater then the other. I named the little fish Sunshine and the big fish Laghpter. The bottom of their tommy is wight. Sunshine is com and Laghpter is jumpie. One is fat and the other is skinny.
Predictions: The snails could eat the fish. The fish could eat the worms. The worms could eat the fish. The fish could eat the snails. The snails could eat the worms. They could die. They could make a home.

Gambusia or a snail. You may want to make clear to students that their "Snail's Diary" should record changes over time, very carefully describe the organism itself and the world as it might find it, and emphasize the way the organism interacts with other organisms and with its environment. (Outcomes 1, 2, 3)

List-an-Adaptation. As a Going Further in Activity 2, older students are asked to make lists of the structures/ behaviors of a particular organism and the benefits of the structure or behavior to that organism. Such a list can assist the teacher in determining how well students understand and can be specific about adaptations. (Outcome 3)

Adding Organisms. A number of possible additions to the habitat are suggested for long-term observation in the classroom. Students predict what might happen when one of these organisms is added to the habitat. Will it be able to survive? Students should explain the reasons for these predictions. Over time, after these organisms are added, teachers can analyze the quality of student descriptions and observations in student journals and class discussions. (Outcomes 1, 2, 3)

Field Trip. During a field trip to an aquatic habitat, the teacher can observe how students incorporate the vocabulary and concepts of the unit into their explorations of a natural environment. Upon arriving at the site, the teacher can ask students to predict some of the organisms they may see, and to describe the similarities and differences between their model and real aquatic habitats. (Outcomes 1, 2, 3)

Literature Connections

At the Edge of the Pond
by Jennifer Owings Dewey
Little, Brown, Boston. 1987
Grades: 1–4

With soft illustrations and poetic text, this book describes life in different areas in and around a pond—at the surface of the pond, along the shoreline, in deep water, and at the bottom of the pond—as well as at different times of the day. In each area and at all times of the day, the pond teems with life.

The Day They Parachuted Cats on Borneo: A Drama of Ecology
by Charlotte Pomerantz; illustrated by Jose Aruego
Young Scott Books/Addison-Wesley, Reading, Massachusetts. 1971
Out-of-print
Grades: 4–7

This cautionary verse, based on a true story, relates how spraying for mosquitoes in Borneo eventually affected the entire food chain, from cockroaches, geckoes, cats, and rats to the river and the farmer. This story illustrates the possible negative consequences of human intervention. The strong, humorous text makes the book a success whether read out loud or performed as a play. The explanation of the food chain makes a nice connection to *Aquatic Habitats*.

An Elephant Never Forgets Its Snorkel: How Animals Survive Without Tools and Gadgets
by Lisa Gollin Evans; illustrated by Diane de Groat
Crown Publishers, New York. 1992
Grades: 3–6

Contains eighteen analogies between human and animal behavior, showing how animals use their bodies in place of the tools, gadgets, and equipment on which humans depend. For example, humans must use a tall ladder to pick fruit from a tree, but giraffes, with their long legs and neck, can graze treetops with ease. A great book about the adaptations of a variety of animals, a subject students can begin to explore through *Aquatic Habitats*.

Hatchet
by Gary Paulsen
Bradbury Press/Macmillan, New York. 1987
Puffin/Viking Penguin, New York. 1988
Grades: 6–12

After a plane crash, 13-year-old Brian must survive alone in the Canadian wilderness. He slowly learns how to provide shelter, fire, and food for himself. He follows some birds to a bush full of berries and learns how valuable it is to observe the animals around him and adapt to his new environment. In several parts of the book Brian is plagued with clouds of mosquitoes and devises ways to avoid them. He also learns about fish as he works to catch them for food.

Incognito Mosquito Makes History
by E. A. Hass; illustrated by Don Madden
Random House, New York. 1987
Grades: 4–7

In this book, the famous insective travels back in time to solve five mysteries involving such notables as Christopher Columbug, Benetick Arnold, Buffalo Bill Cootie, Tutankhamant, and Robin Hoodlum.

Incognito Mosquito, Private Insective
by E. A. Hass
Lothrop, Lee & Shepard Books, New York. 1982
Grades: 4–7

In this book, the first of several, the mosquito detective tells a cub reporter of his exploits and encounters with such insect notables as Mickey Mantis, F. Flea Bailey, and the Warden of Sting Sting Prison. In each chapter, the detective tells of a past case whose solution is at first left for the reader to solve; the final page of the chapter then gives the solution.

Incognito Mosquito Takes to the Air
by E. A. Hass; illustrated by Don Madden
Random House, New York. 1986
Grades: 4–7

While appearing on a TV talk show, the famous insect detective describes his adventures outwitting malefactors and solves a mystery on the air. A very humorous book absolutely overflowing with subtle puns.

Julie
by Jean Craighead George; illustrated by Wendell Minor
HarperCollins, New York. 1994
Grades: 4–Adult

As the sequel to *Julie of the Wolves*, this book continues the saga of
Julie, a brave and wise Eskimo young woman who lives with her
father and his new wife on the North Slope of Alaska. For the
benefit of the whole village, Julie must protect a captive herd of
musk oxen from her pack of wolves. This book is rich with
information about the Arctic ecosystem. Adaptations are frequently mentioned, and the interactions and interdependence of
the Arctic animals are often emphasized. Julie recalls an elder's
words: "We are all here for each other; the Eskimos, the mammals, the river, the ice, the sun, plants, birds, and fish." Other
strong messages in the book are that the habitat provides for the
needs of animals and humans; careful observation leads to understanding; and populations of animals are controlled by other
animals and by humans. With messages like these, this book is
an excellent literature connection for *Aquatic Habitats*.

The Missing 'Gator of Gumbo Limbo: An Ecological Mystery
by Jean Craighead George
HarperCollins, New York. 1992
Grades: 4–7

Sixth-grader Liza K and her mother live in a tent in the Florida
Everglades. She becomes a nature detective while searching for
Dajun, a giant alligator who plays a part in a waterhole's oxygen-
algae cycle, yet is marked for extinction by local officials. The
book is full of detail about the region's flora and fauna and its
interaction with humans. In her forward to the book, the author
states that human beings did not weave the web of life, but are a
strand in it. "Now that we know what we have done to the web,
we see that our role as an intelligent animal is to mend it. There
are millions of Gumbo Limbo Holes on this earth, from a city
window box or vacant lot to streams and lakes to the wilderness
areas of Alaska."

The Old Ladies Who Liked Cats
by Carol Greene; illustrated by Loretta Krupinski
HarperCollins, New York. 1991
Grades: K–6

When the old ladies are no longer allowed to let their cats out at
night, the delicate balance of their island ecology is disturbed, with
disastrous results. Based on Charles Darwin's story about clover
and cats, this ecological folk tale demonstrates the interrelationships
of plants and animals. This book could prompt a discussion of
interrelationships in the classroom aquatic habitats.

Pond Year
by Kathryn Lasky; illustrated by Mike Bostock
Candlewick Press, Cambridge, Massachusetts. 1995
Grades: K–3

This is the delightful story of two girls (best friends who call each
other "scum chums") who love exploring a backyard pond. Written
from the point of view of one of the girls, the story tells the changes
that occur in the pond throughout the year. As the girls observe and
interact with many different plants and animals, science facts are
imparted to the reader. A great book to foster an appreciation for the
diversity of life in a pond, and the value of exploring a natural
environment. Students can compare the changes described in the
pond with the changes in their own "desktop" ponds.

The Salamander Room
by Anne Mazer; illustrated by Steve Johnson
Alfred A. Knopf, New York. 1991
Grades: K–3

A little boy finds an orange salamander in the woods and thinks of
the many things he can do to turn his bedroom into a perfect salamander home. In the process, the habitat requirements of a forest
floor dweller are nicely described.

Why Mosquitoes Buzz in People's Ears: A West African Tale
retold by Verna Aardema; illustrated by Leo and Diane Dillon
Dial Press, New York. 1975
Grades: K–6

This retold West African folk tale cleverly explains why mosquitoes
buzz in people's ears, and how the owl's call is what makes the sun
rise each morning. In a folk-tale sort of way, the interconnectedness
of animals and the disastrous consequences of a simple action are
illuminated in this story.

Summary Outlines

Activity 1: Creating an Aquatic Habitat

Getting Ready

Before the Day of the Activity
1. Decide how to assign students into groups of four and where the habitats will be kept during the unit.
2. Purchase water plants, sand, gravel, and a dechlorinating liquid.
3. Set up your holding tank and fill an additional bucket or other large container with extra dechlorinated water.
4. Prepare the tanks for the student groups.
5. Collect plastic containers to use as shelters in the aquariums.
6. Duplicate four copies of the Aquatic Habitats student sheet for each student to use in the first four activities. If you prefer to use journals with blank pages, decide on materials for your students to use in making their journals or folders.
7. Make one copy of the Aquatic Habitats Task Cards sheet for each student group. Cut and stack them.

On the Day of the Activity
1. Fill a cup with sand and a cup with gravel for each group.
2. Arrange group materials (sand, gravel, Elodea, shelter, tank filled with water, and label) away from student work areas. Set out the Elodea as individual sprigs.
3. Put newspapers or paper towels nearby in case of spills.
4. Have handy the sets of task cards for each team.
5. Prepare a piece of chart paper with the title "Aquatic Habitats Questions," and have it handy to put on the wall.
6. Have the Aquatic Habitats student sheets or journals on hand. Decide whether you want to have students construct the journals before the unit or during the first activity.

Introducing Aquatic Habitats
1. Start by asking, "What kinds of animals and plants live under water?" Accept any ocean or fresh water organisms as answers.
2. Show students one of the tanks and ask what might be able to live in it.
3. Once students have mentioned a variety of fresh water organisms, including fish, ask them to help you think of what a fish would need to live. List their ideas on the board.
4. Write **habitat** on the board, and explain that a habitat is a place that has everything an animal needs to live. Write **aquatic** in front of habitat, and explain that it comes from the Latin word *aqua,* which means water.

5. Tell students that they'll get to set up small aquatic habitats, and will later add small animals to them.
6. Explain how and why you removed the chlorine from the water in their tanks.

Making an Aquatic Habitat

1. Explain that students will work in groups of four to set up and share one aquatic habitat. Put students in their groups and have them number off one to four. Explain that each group member will have a task that goes with their number.
2. Show the materials and explain the four tasks students will follow to change their tanks into aquatic habitats.
3. Pass out the task cards. Say students should do the tasks in order, and cooperate with their teammates.
4. Have students clear their desks, and have each group member get the material for their task while you deliver the tanks full of water and the lids.
5. Circulate among the groups, answering questions and encouraging discussion.
6. Once the habitats are set up, have the teams add lids. Collect empty cups, and mop up any spills.

Documenting the Habitat

1. Let students know that over the next few weeks their aquatic habitats will become "model ponds," and they will be scientists studying things about pond life that people seldom see.
2. As scientists, they need to keep good records. Show students an Aquatic Habitats student sheet.
3. Quickly sketch a sample of the sheet on the board and demonstrate how to fill out the sheet. Let students know that **observations** tell what the tank is like now, and **predictions** are guesses about what the tank might be like in a day or two.
4. Distribute the Aquatic Habitats student sheets or journals. Ask students to write all the observations and predictions they can on their own papers.
5. On the chart paper, start a class list of questions that come up during the investigations.

Activity 2: Tubifex Worms and Snails

Getting Ready

Before the Day of the Activity
1. Obtain snails and Tubifex worms. Keep the snails in the holding tank and the worms in a container with a *small* amount of dechlorinated water.
2. If you intend to use hand lenses, plan some additional class time to give students experience with them.

On the Day of the Activity

1. Prepare two cups of worms and two cups of snails for each student group. Cover the cups with newspaper or otherwise keep them out of sight.
2. Have a sheet of scratch paper, the student sheets or journals, and, if you're using them, a hand lens ready for each pair of students.

Introducing the Activity

1. If you collected the tanks after the previous activity, distribute them to the teams of students.
2. Invite students to share what they've noticed in their model habitats.
3. Remind students that aquatic animals can be injured by chemicals in the water. Ask how they can make sure aquatic animals remain healthy. Mention and, if desired, post the "Habitat Health Rules."

Observing Tubifex Worms

1. Tell students they'll first be observing a common animal similar to earthworms. Write "Tubifex Worm" on the board. Display a cup of worms and let students know that pairs will each receive a cup of worms.
2. Sidestep any negative comments and point out that these creatures have an important role to play in the habitats.
3. Make clear that partners will not put their worms into their tanks right away, but should first put the cup on the table where both students can easily see it, holding the white scratch paper behind the cup to increase visibility of the worms. Encourage specific observations.
4. Distribute the cups and scratch paper. Circulate, reminding students to use their "quiet observer" skills and not to release the worms into the tanks yet.
5. Regain the attention of the class. Have volunteers share their observations. Encourage predictions about what will happen when the worms are added to the habitat.
6. Model how students should gently lower the cup into the water so all the worms can swim out into the tank.
7. Tell students that one pair will release their worms while the whole team observes the behavior of the worms. After a few minutes, the second pair may release their worms.

Discussing Observations and Introducing Water Snails

1. After they have had a few minutes to observe, regain the attention of the class, and invite volunteers to share their observations and questions.
2. Ask questions about the worms, such as how they move, what substrate they seem to prefer, and what they might eat. Point out

that students may be able to discover the answers to these questions by observing the worms in the model habitats.

3. Say that it's time to observe another animal. Show students the snails in cups and invite predictions about what the snails might do when released in the habitat.

4. Challenge students to observe the snails in the cups first, then release them into their tanks. Make sure students understand that these snails need to stay in water to survive.

5. Show students how to gently slide the snails off the surface of the cup and onto the surface of the water habitat.

6. Distribute the cups and ask students to observe the snails. Circulate among the teams listening to observations. For older students (fifth and sixth grade), explain that the behaviors and body structures that help an animal survive in a habitat are called **adaptations.** Review some of the adaptations of snails and Tubifex worms.

7. Have students take out their journals, or provide a new Aquatic Habitats student sheet. Ask students to draw and label the animals in their habitats, and write down their observations and predictions.

8. Collect empty cups, scratch paper, and wipe up any spills.

Activity 3: Fish Enter the Habitat

Getting Ready

Before the Day of the Activity

1. Decide what kind of fish you will use and arrange for their purchase and delivery. Gambusia, goldfish, guppies, and fathead minnows are recommended.

2. Try to schedule the arrival of your fish for a few days before they'll be distributed to students. Add the fish to the holding tank after equalizing the temperature of the water in the transfer container. Feed the fish sparingly, but not immediately before the classroom activity.

On the Day of the Activity

1. For each pair of students put two fish in a cup half-filled with dechlorinated water. Cover the cups with newspaper and set them aside.

2. If you have decided to use them, make a copy of the Parts of a Fish student sheet for each student.

3. Have a sheet of scratch paper, the student sheets or journals, and, if you're using them, a hand lens ready for each pair of students.

Observing Fish

1. If you collected the tanks after the previous activity, distribute them to the teams of students.

2. Write "Gambusia" on the board. Explain that this is the name of the fish they are about to observe and add to their habitats. Tell students that each pair will get a cup with two fish. Make clear that they should not add the fish to the tank until you tell them it's time.

3. Ask students to put the cup where both students can easily see, with a piece of scratch paper under the cup to increase visibility of the fish. Encourage quiet observation of parts of the fish, their colors and behaviors.

4. Distribute the cups and scratch paper.

5. Circulate among students. After they have had time to observe and discuss their fish, regain the attention of the class and ask what they observed.

Fish Into the Habitat

1. Ask for predictions about what will happen when the fish are added to the tank.

2. Model how to gently add the fish to the habitat. Say that one pair will release its fish while the team observes. After a few minutes, the second pair may release its fish.

3. Allow as much time as possible for observation of the fish, as well as observation of any changes in the habitats.

4. Have students take out their journals or student sheets, and allow time for students to draw their tanks and to write down their observations and predictions.

5. Collect empty cups and scratch paper, and wipe up any spills.

Sharing Observations

1. Regain attention of the class. Invite volunteers to tell what happened when they released the fish.

2. Ask about other things they noticed in their model habitats. Record their observations on the chalkboard.

3. When students mention certain body structures or behaviors, ask how they may help the animals survive in an aquatic habitat.

4. Help students understand that each plant or animal has different structures that serve different functions in growth, survival, and reproduction. For older students, emphasize the concept of adaptation—the idea that certain behaviors or body structures can be key to survival.

Predictions and Questions

1. Ask for predictions of what might change in the habitats in the next few days. If students have questions, list them on the chart paper.

2. Ask, "How do fish breathe?" Have students first focus on how we breathe. After having students stand and slowly inhale and exhale a big breath, explain that our lungs take oxygen that we need from the air we breathe in.

3. Say fish need oxygen too, but they do not breathe air with lungs. Instead, they make water flow over body structures called **gills.** The gills take in oxygen from the water. Ask students where they think the gills are located.

4. With student help, sketch and label the parts of a fish on the board. If you've decided to use them, distribute the Parts of a Fish student sheet for students to label, or have students draw and label their own.

5. Encourage students to record more observations and drawings on the student sheets or in their journals.

Activity 4: Mosquitoes

Getting Ready

Before the Day of the Activity
1. Several weeks before the activity, order mosquito eggs or larvae, or create your own mosquito culture.
2. For your extra mosquito larvae, create a separate holding tank using a wide-mouthed container covered with a length of pantyhose.
3. Read the story "Too Many Mosquitoes!" to determine whether and how you will use it with your class.

On the Day of the Activity
1. Prepare a cup of mosquito larvae/pupae for each student.
2. Duplicate a copy of the Mosquito Life Cycle student sheet for each student.
3. Have a sheet of scratch paper, the student sheets or journals, and, if you're using them, a hand lens ready for each student.

Sharing Observations of Changing Habitats
1. If you collected the tanks after the previous activity, distribute them to the teams of students.
2. Begin with an invitation for volunteers to share observations about their changing habitats. Have them refer to their previous student sheets or journals.
3. Ask if students have noticed anything that has floated to the bottom of their habitats. Ask what they think happens to this waste material.
4. Point out that decaying waste is very important in habitats. The Tubifex worms and bacteria help decompose (break down) the decaying waste into nutrients for plants to use.

5. Ask if anyone has noticed algae in their tanks. Emphasize that their Elodea and algae need sunlight, air, and the nutrients made available by decomposers breaking down decaying matter.
6. Review the class list of Aquatic Habitat Questions. Add new questions and answer others.

Observing Mosquito Larvae and Pupae in Cups
1. Ask students if they have ever been bitten by a mosquito, what they know about mosquitoes, and if there are any in the area.
2. Explain that before they become adults, young mosquitoes live in aquatic habitats, and are found in almost every fresh water habitat in the world.
3. Tell the class that each student will get a cup with one or two young mosquitoes to observe and sketch. Ask students to compare their insects with others in their group. Tell students not to add the mosquitoes to their habitats until you tell them to do so.
4. Distribute the cups of larvae/pupae and scratch paper.
5. After students have had time to observe, regain the attention of the class. Ask students to describe and compare their larvae.
6. As students share observations about how the larvae and pupae move, ask how they think those movements help them survive.
7. Explain the difference between the movement patterns of the larvae (wigglers) and the pupae (tumblers). Write these words on the chalkboard and give their singular forms (larva and pupa).
8. Briefly explain the mosquito life cycle.

Adding Mosquito Larvae and Pupae to Aquatic Habitats
1. Ask students for predictions about what might happen when the larvae/pupae are added to their model habitats.
2. Say that each group will pour two cups in their tank, but keep the rest of the larvae and pupae to watch them turn into adult mosquitoes.
3. Review and model the procedure for having students add two cups of mosquitoes to their tanks and for returning two cups to the area of the mosquito holding tank. Remind teams to put the lids on their tanks when they're finished adding their larvae.
4. Circulate among the groups. Remind students to act like scientists and use quiet observation techniques.
5. Have students take out their journals or student sheets, and allow time for drawing their habitats and writing observations and predictions. Also have students record some of the other events they observed.
6. While students are working, distribute the Mosquito Life Cycle student sheets.
7. Transfer the "saved" larvae from the cups into the mosquito holding tank.

8. Collect extra cups, and wipe up any spills.

9. Provide time in the next few days/weeks for continued observations and predictions about the aquatic habitats and the mosquito holding tank, with continued journal writing and record-keeping.

Discussing Observations and Introducing Food Webs

1. Gather students away from their habitats for the discussion. Invite them to share observations.

2. Reveal the other name for Gambusia—mosquito fish! Say that mosquito fish are added to lakes and ponds to control the population of mosquitoes so poisons don't have to be used. Introduce and define **biological control** as the name given to any organism used to control the population of another organism.

3. Ask students what each organism in their tank eats. Sketch the organisms on the chalkboard and draw arrows to show the direction nutrients go.

4. Explain that if fish die—or as their feces or waste falls to the bottom—decomposers like Tubifex worms and tiny bacteria break them down to release nutrients for plants. Draw an arrow from a fish to the Tubifex worms.

5. Explain that the same is true for snails, and draw an arrow from snails to the worms.

6. Have older students help you draw an example of a nutrient cycle. Also, extend the food web beyond the classroom habitats.

7. To emphasize that algae, Elodea, and all plants need sunlight, draw the sun shining on the chalkboard food web.

8. You could have students draw some food webs. Students could sequence just two organisms at a time or create whole cycles or webs. Encourage students to create drawings of food webs in other habitats.

Activity 5: Further Explorations

Part I: Ongoing Investigations with Classroom Aquatic Habitats

Recording New Developments

1. Over a period of several weeks, have students document observations with journal entries and drawings.

2. Encourage students to think about questions they have and further investigations they may want to make.

Designing and Conducting Experiments

Based on their own observations and questions, and with your feedback and suggestions, ask students to come up with a question to investigate.

Decomposition
1. If a fish has died, students may be interested in observing the process of decay and making a closer study of the process of decomposition.
2. To minimize smells, put the dead fish in a container with some aquarium water, then cover with a slightly perforated lid. Place in a well ventilated location.
3. Students may also want to experiment to see if Tubifex worms will consume the body.

Observing Changes as Habitats are Dismantled
1. When you near the end of the unit, provide time for students to disassemble the aquariums.
2. As the groups work together to return the fish, snails, worms, and Elodea to the classroom holding tank, encourage everyone to look for changes in their tanks.
3. Have students write a closing essay or summation in their journals.

Part II: An Aquatic Field Trip

Getting Ready

Before the Day of the Field Trip
1. Visit the site yourself so you can better plan for student's investigations and safety. Make certain you know the regulations of the site and are aware of any endangered species.
2. Arrange for several adult volunteers to come along. Make sure they are clear on their roles and know what is expected of students.

On the Day of the Field Trip
1. In the classroom, go over guidelines with your students to help them gain respect for the habitat.
2. Discuss safety issues with students, including all necessary precautions to be taken around water. You may also want to discuss what they should wear and what to bring and not bring along on the trip.

Sensing the Beauty of the Habitat
Begin the field activities *away* from the water's edge with a three to five-minute, whole-class sensory observation.

Introducing the Procedures
1. Away from the water's edge, review the boundaries, safety procedures, and other rules designed to minimize impact on the environment.

2. Let students know that they'll be exploring several parts of the aquatic habitats to find out what lives in them.

3. Tell students the ways to sample habitats and remind them to try many different habitats. When they have samples from two slightly different water habitats, they should compare what they see.

Sampling the Aquatic Habitats and Sharing Findings

1. Distribute the materials to each group and challenge them to find as many different kinds of plants and animals as possible. Point out that one of each kind is enough, and that they should share their findings with other groups.

2. Circulate among the student groups, encouraging teamwork and sharing of discoveries. Allow as much time as possible for the sampling activity.

3. Gather students away from the water's edge and ask them to arrange their observation trays in a large circle so everyone can walk around and look over the contents.

4. Gain the attention of the whole class and ask them to help you list the plants and animals that were found.

5. If time allows, have students resume their sampling with more questions in mind.

6. Have the teams return their organisms to the water where they were found.

Making Connections Back in the Classroom

1. Give students an opportunity to share their discoveries. Ask how the pond was similar to and different from their model habitats.

2. Ask questions to help students understand and better appreciate the complexity and ecological balance of the pond food web.

3. Conclude by having students write about the aquatic field trip in their journals.

Going Further with the Aquatic Field Trip Activities

You and your students may choose to undertake one or more of these further investigations at the field trip site: crayfishing, can fishing, or observing water striders. Or come up with ideas of your own!

Blood-Sucking Friends

Note: In performance, the verses for this song are spoken dramatically, in vampire-style. The chorus and bridge are sung, with some spoken words interspersed. One set of lines in the chorus changes each time, while the rest of the chorus stays the same. For example, the lines "When you give you're giving life/ (We're your friends)/Not for me but for my wife" in the first chorus are replaced by "Us males don't whine or dine/(We're your friends)/Plant juices taste just fine" in the second chorus. For musical arrangement, guitar chords, and a recorded version of this song, contact the Bungee Jumpin' Cows at the address and website shown below.

Prologue, spoken over guitar chord:
You get into bed and turn off the lights
Ready to sleep you say goodnight
Then you hear a buzzing getting near
Before you know it a mosquito's in your ear

Verse 1:
You try to kill it
You cover your head
But mosquito babies must be fed
You smack and you whack
You think you're a stud
But she flies away with a belly full of blood

Chorus (sung):
We're your blood-sucking friends
(We're your blood-sucking friends)
We're your blood-sucking friends
On your blood we depend
(We're your friends)
When you give, you're giving life
(We're your friends)
Not for me, but for my wife
We're your blood-sucking friends

Bridge: (sung)
I feed frogs, snakes, lizards, salamanders and fish
Dragonflies and swallows think that I'm a hot dish
Spiders and water boatmen, backswimmers too
They all feed on me but now I'm gonna feed on you

Verse 2:
The buzz you hear is her love call for a mate
If a male's antennae hears it, he meets her for a date
Us males are quiet and we don't bite
But can you hear that female, she'll drink to you tonight

Chorus:
Us males don't whine or dine
Plant juices taste just fine

Verse 3:
When we bite we leave spit, that's how we spread disease
Civilizations we've brought to their knees
Yellow fever and malaria in tropical lands
All I spread here is itch, but you're still not my fans!

Chorus:
Help out our little ones
Without your blood their life is done

Bridge (with first half of chorus):
Wouldn't you like to donate?
You'll feel just great

Verse 4:
She was born in stagnant water in a raft of eggs
A cute little wriggler without any legs
Living life as a pupa wasn't much of a thrill
But now she's an adult and it's time to drill!

(Chorus sung twice, with lines below interspersed as asides)

You might like us all to die
But don't even try
I think you'd best refrain
You'll mess up the food chain
Please give your blood freely
We accept type A or B
Ve vant to suck your blood!

"Blood-Sucking Friends" is one of many songs on the CD entitled "Rockin' the Foundations of Science," by the Bungee Jumpin' Cows, C.U.D. Productions, 413A 61st Street, Oakland CA 94609. For more information on this and their other science songs and CDs, the Bungee Jumpin' Cows can be reached at (888) 434-COWS or check out their website at: www.moo-boing.com/bjc

Aquatic Habitat Task Cards

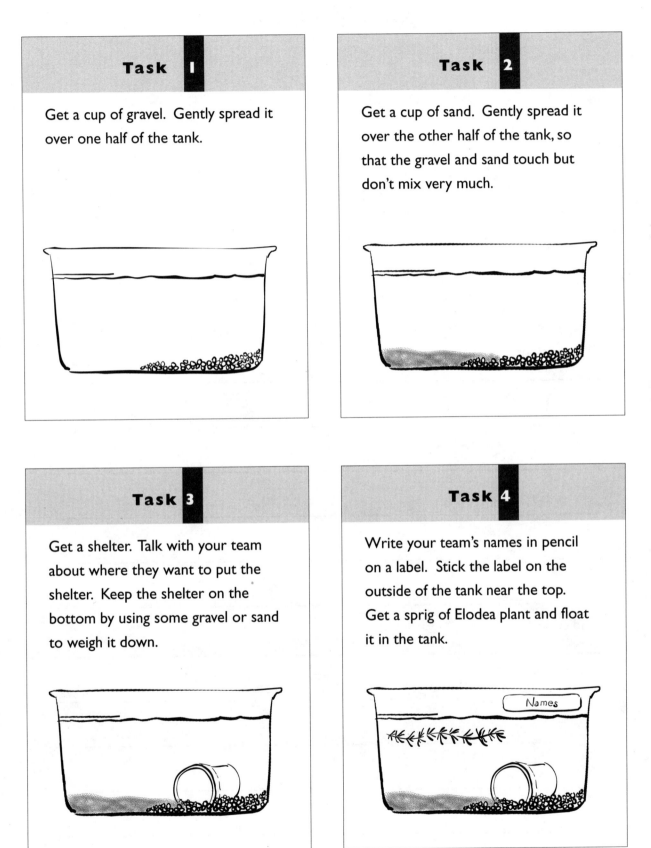

Task 1

Get a cup of gravel. Gently spread it over one half of the tank.

Task 2

Get a cup of sand. Gently spread it over the other half of the tank, so that the gravel and sand touch but don't mix very much.

Task 3

Get a shelter. Talk with your team about where they want to put the shelter. Keep the shelter on the bottom by using some gravel or sand to weigh it down.

Task 4

Write your team's names in pencil on a label. Stick the label on the outside of the tank near the top. Get a sprig of Elodea plant and float it in the tank.

Names

AQUATIC HABITATS

Observations:

_____

_____

_____

_____

_____

Predictions:

_____

_____

_____

_____

Mosquito Life Cycle

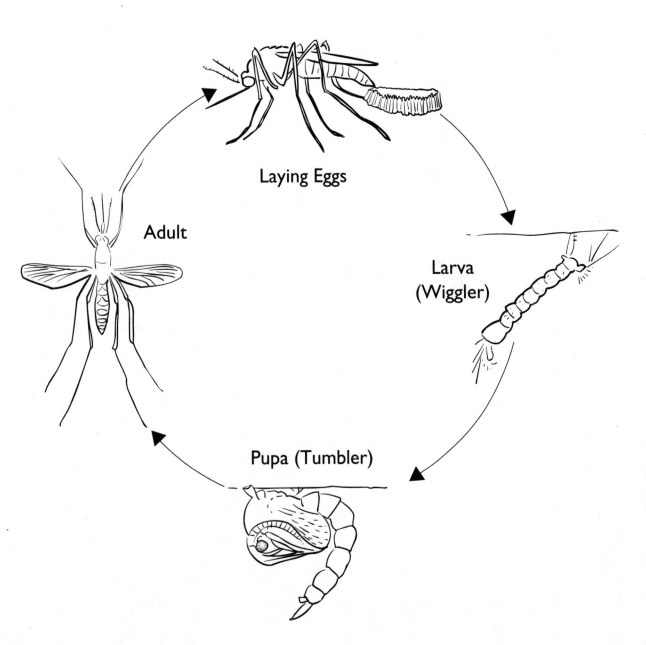

Laying Eggs

Adult

Larva (Wiggler)

Pupa (Tumbler)

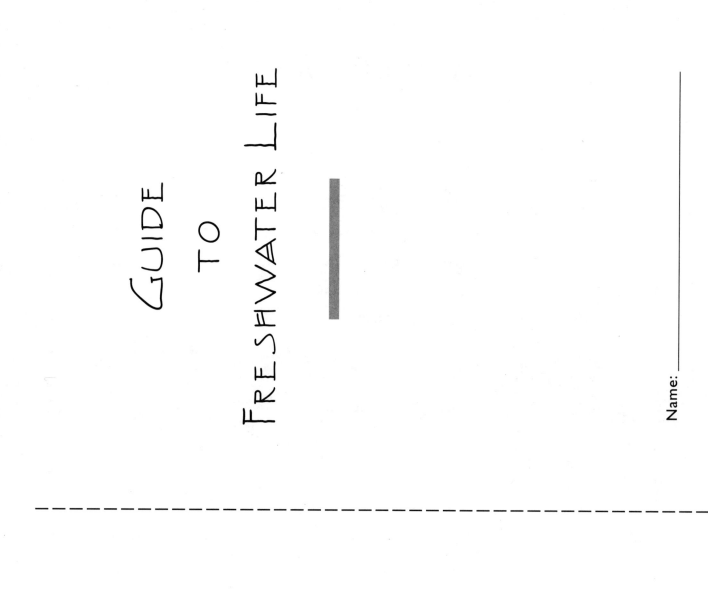

GUIDE TO
FRESHWATER LIFE

Name: _____

Aquatic Animals

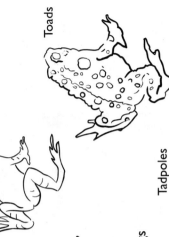

FISH

Fish spend their lives entirely in water, breathing by means of gills. They have fins, and their streamlined bodies are usually covered with scales.

AMPHIBIANS

Amphibians begin life with gills in water and later develop lungs. Their skin is thin, scaleless, smooth or warty, and usually moist. Frogs, toads, and salamanders belong in this group.

Frogs
Frogs are smooth skinned with long, powerful hind legs. Tree frogs have toes with enlarged tips.

Toads
Toads possess a warty skin, large neck glands, and are rarely found moving about during the day. Toads have shorter back legs than frogs have.

Tadpoles
Tadpoles are the well-known larvae of frogs and toads. They are completely aquatic.

Salamanders
Salamanders also include newts. They have lizardlike bodies but lack claws.

Salamander Larvae
These larvae are completely aquatic and possess external gills, which can help you distinguish these larvae from tadpoles.

Toads

Frogs

Tadpoles

Salamanders

Salamander Larvae

Snail Eggs

Snails

Leeches

Tubifex Worms

Flatworms

Clams and Mussels

Copepods

Crayfish

Scuds

Water Fleas

Seed Shrimp

Water Mites

Aquatic Insects

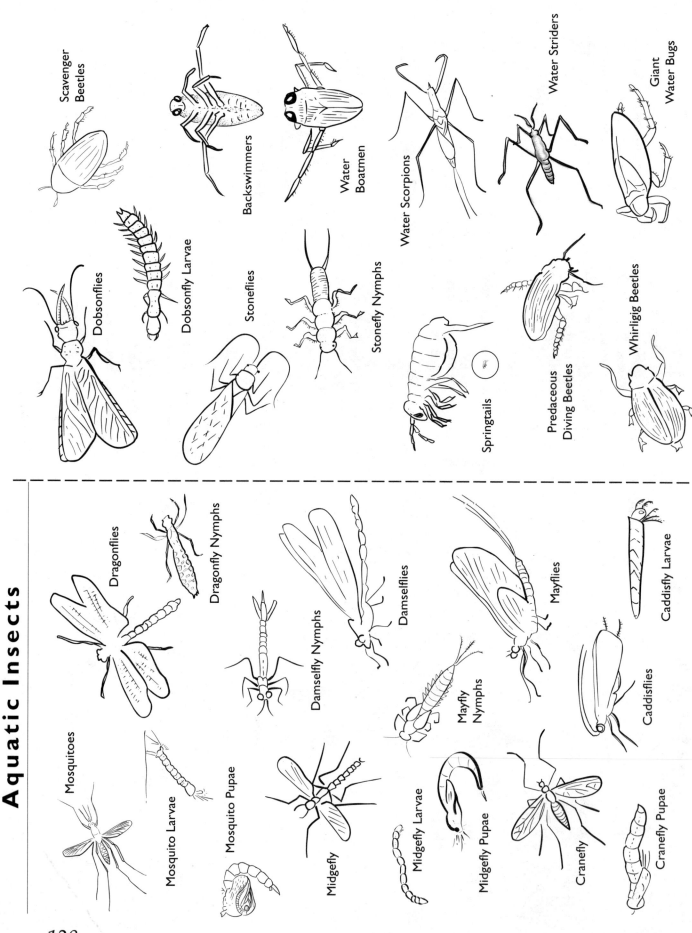

Scavenger Beetles

Backswimmers

Water Boatmen

Water Scorpions

Water Striders

Giant Water Bugs

Dobsonflies

Dobsonfly Larvae

Stoneflies

Stonefly Nymphs

Springtails

Predaceous Diving Beetles

Whirligig Beetles

Dragonflies

Dragonfly Nymphs

Damselfly Nymphs

Damselflies

Mayflies

Caddisfly Larvae

Mayfly Nymphs

Caddisflies

Mosquitoes

Mosquito Larvae

Mosquito Pupae

Midgefly

Midgefly Larvae

Midgefly Pupae

Cranefly

Cranefly Pupae

128